FOURIER TRANSFORMS
AND THEIR
PHYSICAL APPLICATIONS

Techniques of Physics

Editors
G. K. T. Conn and K. R. Coleman

Techniques of physics find wide application in biology, medicine, engineering and technology generally. This series is devoted to techniques which have found and are finding application. The aim is to clarify the principles of each technique, to emphasize and illustrate the applications and to draw attention to new fields of possible employment.

1. D. C. Champeney: Fourier Transforms and their Physical Applications, 1973.

In preparation

K. D. Froome: Electromagnetic Distance Measurement.

J. B. Pendry: Low Energy Electron Diffraction.

R. E. Meads: Mossbauer Techniques.

M. Pepper and W. Eccleston: Physical Principles of MOS Devices.

G. H. Lunn: The Practice of High Speed Photography.

B. R. Garfield: Photo-emissive Surfaces.

FOURIER TRANSFORMS AND THEIR PHYSICAL APPLICATIONS

D. C. CHAMPENEY

School of Mathematics and Physics,
University of East Anglia,
Norwich, England

1973

ACADEMIC PRESS
London and New York

ACADEMIC PRESS INC. (LONDON) LTD.
24/28 Oval Road,
London NW1

United States Edition published by
ACADEMIC PRESS INC.
111 Fifth Avenue
New York, New York 10003

Library of Congress Catalog Card Number: 72–84350
ISBN: 0–12–167450–9

Printed in Great Britain by
ROYSTAN PRINTERS LIMITED
Spencer Court, 7 Chalcot Road
London NW1

Preface

Fourier transforms, with the associated concepts of convolution and correlation, are being increasingly used in all branches of the physical sciences, and this book is intended both as an introduction and as a reference book for those whose work brings them into contact with the subjects, be they advanced undergraduates, postgraduates or others. Whilst the systematic use of Fourier transforms is well established in electrical engineering, their treatment in the other physical sciences often appears only in appendices to books on subjects such as optics, electricity, quantum mechanics, or acoustics. However, once a technique becomes so universally applied as are Fourier transforms there is a need for a unified treatment in which the mathematical methods are both explained and also applied to a range of different phenomena. Subjects such as vectors and wave motion are often treated in this way, and it is this approach that I have attempted in this book.

In the arrangement of the book I have followed the obvious and logical procedure of describing the mathematical techniques in Part I, and the applications in Part II. However, readers new to the subject might benefit by using Parts I and II simultaneously, rather than reading all of Part I before looking at Part II, since it is the applications which bring the equations to life. One-dimensional transforms and spectra might be covered first in Chapters 1, 2 and 4 (omitting Chapter 3) providing a sufficient mathematical grounding for much of the material on linear systems, harmonic oscillators, and electrical circuits and filters in Chapters 7, 8, and 9. Subsequently the reader might either extend the concepts into two or more dimensions using Chapter 3 with the diffraction theory in Chapter 11, or he might follow up the subjects of correlation and random functions in Chapters 5 and 6, together with the theory of information retrieval described in Chapter 10. The remaining Chapters 12, 13 and 14 covering aspects of optical coherence, holography, and X-ray, neutron, and electron diffraction provide further applications which are slightly more advanced.

Since Fourier transforms are applied so widely an exhaustive coverage of applications is clearly impossible and the topics chosen are intended only as typical ones providing close analogies for many others. The subjects are mostly within classical as opposed to quantum physics; this seems appropriate since there are few branches of quantum mechanics which do not possess a close analogy in some part of classical physics, and since on the whole classical descriptions are still more widely understood.

In Part I I have laid emphasis on results rather than proofs since the book is intended for applied scientists rather than mathematicians, and many of the proofs are placed in appendices. For more rigorous treatments I have referred the reader to other appropriate sources. In Part II the descriptions of the applications are intended to show how Fourier transforms may usefully be applied rather than to introduce the subjects themselves, and some previous knowledge of the phenomena is assumed.

I have frequently adopted a concise approach in order to cover the ground in a book of this size and have omitted the more obvious proofs in several places in order to maintain the flow of ideas. Many of the little theorems relating to Fourier transforms are in fact easy to establish (for instance the results of shifting and scaling) and in such cases I have simply drawn attention to the result and have encouraged the reader to seek his own proof with the deliberate use of the phrase 'it can be shown that . . .'.

My interest in this subject has arisen out of experimental work carried out in the School of Mathematics and Physics at the University of East Anglia and it is a pleasure to acknowledge the stimulus and help derived from discussions with my colleagues and students. I wish also to thank my wife for her consistent encouragement and help throughout.

David Champeney,

November 1972

Contents

PART I – MATHEMATICAL FOUNDATIONS

Chapter 1

Fourier Series

1.1 Introduction

It is well known that when a quantity varies periodically with time it may be 'analysed into its harmonic components'. The quantity may be the pressure at a point varying due to the passage of a sound wave past it, it might be the electric field strength at a point varying due to the passage of a light or radio wave past it, or it might be the voltage or current somewhere in an electric circuit. In each case provided the variation repeats itself at some basic frequency then the disturbance can be considered as being built up from a set of harmonically varying disturbances having repetition frequences equal to multiples of the basic frequency. This is so even if the variation within one repetition period is quite complicated and irregular.

This analysis can receive a striking practical demonstration in the case of sound waves. By having a large array of loud speakers each emitting a harmonic wave at one or other multiple of the basic frequency it is possible to build up any desired wave pattern to a close approximation by adjusting the amplitudes and phases of the various emitters. In this way the complicated repetitive wave form produced by a musical instrument can be simulated. The method is limited in practical use since musical sounds are characterised by pressure variations which consist of an aperiodic part (due to hiss and initial transients) superimposed on a repetitive part. The fact that this harmonic analysis and synthesis is possible depends partly on the fact that to a good approximation sound waves in air superimpose to give a net pressure fluctuation which is the algebraic sum of what each wave would

1

produce in the absence of the others, and partly on the purely mathematical results to be described here having to do with Fourier series expansions.

In the examples quoted above time and frequency appeared in each case as a related pair of variables; however further systems exist where the same principles apply except that other pairs of variables are used. For instance the variables distance and inverse wavelength appear in the analysis of the vibratory motion of a stretched string; two vectors related respectively to the position of a scattering element and the direction of the scattered radiation appear in diffraction problems, and two vectors representing position and momentum appear in quantum mechanics.

Most of this book is concerned not with Fourier series expansions, which are appropriate for the analysis of periodic functions, but with Fourier transforms which perform a similar role in the analysis of functions which are not necessarily periodic. The Fourier transforms are thus more general in application, and as we shall see can be considered to reduce to Fourier series in special cases. We start however with Fourier series since they are an important special case and serve as a good (but by no means essential) introduction to the transforms.

1.2 The Basic Formulae for Fourier Series Expansions

If the function $f(t)$ represents the quantity that varies with time, and if τ is the basic repetition period, then we shall call $2\pi/\tau$ the fundamental frequency, ω, of the system expressed as radians per unit time. The statement that $f(t)$ can be analysed into an infinite sum of harmonic components at multiples of the fundamental frequency may be represented mathematically by any of the equivalent eqns (1.1)–(1.5) below, known as Fourier series expansions. It is a matter of convenience or taste which one to use.

$$f(t) = a_0 + \sum_{n=1}^{\infty} \{a_n \cos n\omega t + b_n \sin n\omega t\} \tag{1.1}$$

$$= \sum_{n=0}^{\infty} \{a_n \cos n\omega t + b_n \sin n\omega t\} \tag{1.2}$$

$$= a_0 + \sum_{n=1}^{\infty} c_n \cos (n\omega t + \phi_n) \tag{1.3}$$

$$= a_0 + \sum_{n=1}^{\infty} d_n \sin (n\omega t + \theta_n) \tag{1.4}$$

$$= \sum_{n=-\infty}^{+\infty} g_n e^{+in\omega t}. \tag{1.5}$$

The coefficients a_n, b_n, c_n, d_n, g_n represent the unknown amplitudes and ϕ_n and θ_n the unknown phases. The determination of these coefficients represents the central problem of harmonic analysis, and Fourier discovered that they could be evaluated from a knowledge of $f(t)$ by performing the following integrations. We also include in these equations the various relationships between the coefficients.

$$a_0 = \frac{1}{\tau} \oint f(t)\, dt = g_0 = \text{mean value of } f(t) \tag{1.6}$$

$$a_n = \frac{2}{\tau} \oint f(t) \cos n\omega t\, dt = g_n + g_{-n} = c_n \cos \phi_n = d_n \sin \theta_n \quad [n \geqslant 1] \tag{1.7}$$

$$b_n = \frac{2}{\tau} \oint f(t) \sin n\omega t\, dt = i(g_n - g_{-n}) = -c_n \sin \phi_n = d_n \cos \theta_n \quad [n \geqslant 1] \tag{1.8}$$

$$g_n = \frac{1}{\tau} \oint f(t)\, e^{-in\omega t}\, dt \quad [n = 0,\ \pm 1,\ \pm 2, \ldots]. \tag{1.9}$$

The following further expressions help in relating the coefficients together.

$$g_n = \tfrac{1}{2}(a_n - ib_n) \qquad g_{-n} = \tfrac{1}{2}(a_n + ib_n) \qquad (n \geqslant 1) \tag{1.10}$$

$$c_n{}^2 = d_n{}^2 = a_n{}^2 + b_n{}^2 = 4g_n g_{-n} \quad \text{(sign ambiguity)} \ (n \geqslant 1) \tag{1.11}$$

$$\tan \phi_n = -b_n/a_n \qquad \tan \theta_n = a_n/b_n \quad \text{(phase ambiguity)} \ (n \geqslant 1). \tag{1.12}$$

In the above expressions $\oint$ means an integration over any full period of $f(t)$, that is integration from t_1 to $t_1 + \tau$ where t_1 is arbitrary. Note that in performing these integrations on some given function $f(t)$ onesworking can often be simplified by a judicious choice of the region of integration, i.e. the choice of t_1. It might for instance be simpler to work from $t = -\tau/2$ to $t = +\tau/2$ than from $t = 0$ to $t = \tau$. If we ignore the difficult question of whether such series expansions are possible, but simply assume them to be valid, then the derivation of eqns (1.6) to (1.9) is simple and very elegant and is carried out in Appendix A.

1.3   Useful Information Relating to the Basic Equations

The allowed form of $f(t)$. The function $f(t)$ must stretch from $t = -\infty$ to $t = +\infty$, must repeat itself with some characteristic period τ and must be single valued. $f(t)$ may be a complex function, as also may be the various coefficients. Note that the complex Fourier series, eqn (1.5), is an equality and if $f(t)$ is real we do *not* understand $f(t)$ to be merely the real part of the complex sum; in this case the g_n are suitably complex so that the summation comes out to be real. $f(t)$ can contain discontinuities (as in a square, or saw-tooth waveform), and will be a suitable function if it has only a number of maxima or minima per period, and (c) has no infinite discontinuities. These conditions are known as Dirichlet's conditions and are discussed more fully for instance in references 9, 16, 20 and 21. In addition $f(t)$ may contain a finite number of delta functions per period. Delta functions are discussed in Appendix B.

Simplifying Conditions. If $f(t)$ is an even function $\left(f(t) = f(-t) \right)$ then only cosine terms are needed in eqn (1.2), and if $f(t)$ is an odd function $\left(f(t) = -f(-t) \right)$ then only the sine terms are needed. If the position of the origin is not important then it is possible with some functions to choose the origin so that the function becomes odd or even and thus to simplify one's working. Other simplifications occur if $f(t)$ is either real or pure imaginary. The results are summarised below.

If $f(t)$ is real then: $a_n, b_n, c_n, d_n, \phi_n, \theta_n$ are real

g_n is complex

$g_n = g^*_{-n}$ $a_n = \mathrm{Re}\ 2g_n$

$g_n^* = g_{-n}$ $b_n = -\mathrm{Im}\ 2g_n$.

If $f(t)$ is pure imaginary then: a_n, b_n, c_n, d_n are pure imaginary

ϕ_n, θ_n are real

g_n is complex

$g_n = -g^*_{-n}$ $a_n = 2i\,\mathrm{Im}\,g_n$ $[n \geqslant 1]$

$g_n^* = -g_{-n}$ $b_n = 2i\,\mathrm{Re}\,g_n$.

If $f(t)$ is even then: $g_n = g_{-n}$ $\phi_n = 0$ $\theta_n = \pi/2$ $a_0 = g_0$

$b_n = 0$ $a_n = c_n = d_n = 2g_n$ $[n \geqslant 1]$.

If $f(t)$ is odd then: $g_n = -g_{-n}$ $\phi_n = -\pi/2$ $\theta_n = 0$

$a_n = 0$ $b_n = c_n = d_n = 2ig_n$.

If $f(t)$ is real and even
            or $\Big\}$ *then*: g_n is real.
pure imaginary and odd

If $f(t)$ is real and odd
            or $\Big\}$ *then*: g_n is pure imaginary.
pure imaginary and even

Complex phase angles. The expansions involving θ_n or ϕ_n are inconvenient if θ_n and ϕ_n come to be complex, which means that the expansions remain convenient if $f(t)$ is either real, pure imaginary, even, or odd. The eqns (1.1)–(1.12) do remain valid in all cases however and can be interpreted consistently by means of the relations $\sin ix = i \sinh x$, $\cos ix = \cosh x$, $\tan ix = \tanh x$. It should be noted that in obtaining values of c_n, d_n, ϕ_n and θ_n from a_n and b_n the eqns (1.11) and (1.12) leave ambiguities of sign and phase which do not arise if eqns (1.7)–(1.8) are used.

Half wave series. Although an aperiodic function cannot be represented by a Fourier series, it is possible to derive a series expansion which is valid over a limited range, say from $t = 0$ to $t = T$. This can be achieved by first constructing some periodic function which equals $f(t)$ over the region of interest (but not necessarily outside it) and then finding a series representation for this periodic function. The period τ of the repetitive function may clearly have any value greater than T. If $\tau \geqslant 2T$ one may moreover represent $f(t)$ over the region of interest using an expansion involving only sine or cosine terms. For the special choice $\tau = 2T$ such series are called half wave sine or cosine series. Putting $\omega = 2\pi/\tau = \pi/T$ the following results are obtained:

$$f(t) = a_0 + \sum_{n=1}^{\infty} a_n \cos n\omega t = a_0 + \sum_{n=1}^{\infty} b_n \sin n\omega t \quad \text{(valid } 0 < t < T) \quad (1.13)$$

$$a_0 = \frac{1}{T} \int_0^T f(t)\, dt \quad (1.14)$$

$$a_n = \frac{2}{T} \int_0^T f(t) \cos n\omega t\, dt \quad [n \geqslant 1] \quad (1.15)$$

$$b_n = \frac{2}{T} \int_0^T f(t) \sin n\omega t\, dt. \quad [n \geqslant 1] \quad (1.16)$$

The ways in which the sine and cosine series match $f(t)$ over the region of interest for a simple ramp function are shown in Fig. 1.1.

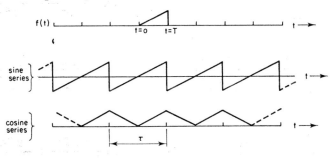

FIGURE 1.1

Gibbs' Phenomenon. Although a function containing a finite step can be synthesized using a Fourier series, it is found that if only a finite number of terms in the expansion are used then the synthesized function overshoots at the ends of the step and then oscillates. Moreover as the number of terms used tends to infinity the amount of the overshoot does *not* tend to zero, though the period of the oscillations does tend to zero. For many purposes this is of no consequence since the average value of the synthesized function over a finite range of t can be made to match the actual function to any desired accuracy by choosing enough terms. It does mean that care is necessary in differentiating a series expansion in such cases. The phenomenon is discussed more fully for instance in references 16 and 20.

1.4 Table of Fourier Series for a Few Illustrative Functions

$$\frac{A}{2}+\frac{2A}{\pi}\left\{\cos \omega t-\frac{\cos 3\omega t}{3}+\frac{\cos 5\omega t}{5}-\cdots\right\}$$

$$\frac{2A}{\pi}\left\{\sin \omega t+\frac{\sin 3\omega t}{3}+\frac{\sin 5\omega t}{5}+\cdots\right\}$$

$$\frac{4A}{\pi^2}\left\{\cos \omega t+\frac{\cos 3\omega t}{3^2}+\frac{\cos 5\omega t}{5^2}+\cdots\right\}$$

$$\frac{4A}{\pi^2}\left\{\sin \omega t-\frac{\sin 3\omega t}{3^2}+\frac{\sin 5\omega t}{5^2}-\cdots\right\}$$

$$\frac{A}{\pi}\left\{\sin \omega t-\frac{\sin 2\omega t}{2}+\frac{\sin 3\omega t}{3}-\cdots\right\}$$

$$-\frac{A}{\pi}\left\{\sin \omega t+\frac{\sin 2\omega t}{2}+\frac{\sin 3\omega t}{3}+\cdots\right\}$$

Half wave rectified cosine wave

$$\frac{A}{\pi} + \frac{A}{2}\cos \omega t + \sum_{n=1}^{\infty} (-1)^{n+1}\frac{2A\cos 2n\omega t}{\pi(4n^2 - 1)} \quad (1.23)$$

Half wave rectified sine wave

$$\frac{A}{\pi} + \frac{A}{2}\sin \omega t - \sum_{n=1}^{\infty}\frac{2A\cos 2n\omega t}{\pi(4n^2 - 1)} \quad (1.24)$$

Equally spaced delta functions
of unit area

$$\frac{2}{\tau}\{\tfrac{1}{2} + \cos \omega t + \cos 2\omega t + \cos 3\omega t + \ldots \} \quad (1.25)$$

or

$$\frac{1}{\tau}\sum_{n=-\infty}^{+\infty} e^{in\omega t}$$

Chapter 2

Fourier Transforms

2.1 Introduction

Whereas the Fourier series allows a periodic function to be represented as an infinite sum of harmonic oscillations at definite frequencies equal to multiples of the fundamental, so the Fourier Transform allows an aperiodic function to be expressed as an integral sum over a continuous range of frequencies. As in Section 1.1 the acoustic realization of such a synthesis depends partly on the physical factor of whether or not acoustic waves superpose linearly to give a resultant pressure equal to the algebraic sum of the individual pressures that each would produce in the absence of the others, and partly on the mathematical results described in the rest of this chapter.

Although we have considered aperiodic functions above, in order to contrast integral and series representations, it will appear subsequently that with a suitable use of delta functions the Fourier transform may be used to cover both periodic and aperiodic functions. The Fourier series then comes to be regarded as a special case of the Fourier transform.

2.2 The Basic Formulae for Fourier Transforms

If $f(t)$ represents a quantity that varies with time, the statement that $f(t)$ can be analysed into an integral sum of harmonic oscillations over a continuous range of frequencies may be represented mathematically by any of the equivalent eqns (2.1)–(2.6) below known as Fourier integrals. It is a matter of convenience, taste, or convention which we use.

$$f(t) = \int_0^\infty \{A(\omega)\cos\omega t + B(\omega)\sin\omega t\}\, d\omega \tag{2.1}$$

$$= \int_0^\infty C(\omega)\cos(\omega t + \phi(\omega))\, d\omega \tag{2.2}$$

$$= \int_0^\infty D(\omega)\sin(\omega t + \theta(\omega))\, d\omega \tag{2.3}$$

$$= \frac{1}{\sqrt{2\pi}}\int_{-\infty}^{+\infty} E(\omega)\, e^{+i\omega t}\, d\omega \tag{2.4}$$

$$= \frac{1}{2\pi}\int_{-\infty}^{+\infty} F(\omega)\, e^{+i\omega t}\, d\omega \tag{2.5}$$

$$= \int_{-\infty}^{+\infty} G(\nu)\, e^{+i2\pi\nu t}\, d\nu. \tag{2.6}$$

The determination of the functions $A(\omega)$, $B(\omega)$, ... $F(\omega)$, $G(\nu)$ represents the central problem of Fourier analysis. Each of the functions $E(\omega)$, $F(\omega)$ and $G(\omega)$ is known as the Fourier transform of $f(t)$, there being no universally accepted convention as to which is meant by this title. In the above equations, ω represents an angular frequency in radians per second and ν a frequency in cycles per second. The functions $A(\omega)$ etc can be found by performing the integrals below, as is shown in Appendix C. We also include in the following equations the various interrelations between the coefficients.

$$A(\omega) = \frac{1}{\pi}\int_{-\infty}^{+\infty} f(t)\cos\omega t\, dt = \frac{1}{2\pi}\{F(\omega) + F(-\omega)\}$$
$$= C(\omega)\cos\{\phi(\omega)\} = D(\omega)\sin\{\theta(\omega)\} \tag{2.7}$$

$$B(\omega) = \frac{1}{\pi}\int_{-\infty}^{+\infty} f(t)\sin\omega t\, dt = \frac{i}{2\pi}\{F(\omega) - F(-\omega)\}$$
$$= -C(\omega)\sin\{\phi(\omega)\} = D(\omega)\cos\{\theta(\omega)\} \tag{2.8}$$

$$E(\omega) = \frac{1}{\sqrt{2\pi}}\int_{-\infty}^{+\infty} f(t)\, e^{-i\omega t}\, dt = \frac{1}{\sqrt{2\pi}}\, G(\omega/2\pi) \tag{2.9}$$

$$F(\omega) = \int_{-\infty}^{+\infty} f(t)\, e^{-i\omega t}\, dt$$

$$= \pi\{A(\omega) - iB(\omega)\} \qquad \text{[for } \omega > 0\text{]}$$

$$= \pi\{A(|\omega|) + iB(|\omega|)\} \qquad \text{[for } \omega < 0\text{]} \tag{2.10}$$

$$G(v) = \int_{-\infty}^{+\infty} f(t)\, e^{-i2\pi vt}\, dt = F(2\pi v). \tag{2.11}$$

The following further expressions help in relating the various coefficients together.

$$C^2(\omega) = D^2(\omega) = A^2(\omega) + B^2(\omega) = \frac{1}{\pi^2}\{F(\omega)F(-\omega)\} \tag{2.12}$$

$$\tan\{\phi(\omega)\} = -B(\omega)/A(\omega). \qquad \tan\{\theta(\omega)\} = A(\omega)/B(\omega). \tag{2.13}$$

2.3   Useful Information Relating to Fourier Integrals

Nomenclature and definition. There is no universally accepted convention governing the definitions of the terms 'Fourier transform' and 'inverse transform'. We shall adopt the notation below, using now the variables x and y instead of t and ω for purposes of generality.

$$f(x) = \frac{1}{2\pi}\int_{-\infty}^{+\infty} F(y)\, e^{+ixy}\, dy = \text{Inverse transform of } F(y) = FT^{+}\{F(y)\}$$

$$F(y) = \int_{-\infty}^{+\infty} f(x)\, e^{-ixy}\, dx = \text{Fourier transform of } f(x) = FT^{-}\{f(x)\}$$

$$f(x) \leftrightarrow F(y). \tag{2.14}$$

This definition results in a lack of symmetry since a factor $(1/2\pi)$ appears in the expression for the inverse transform which is not present in the Fourier transform itself. Symmetry in this respect may be achieved if a definition is chosen which incorporates a factor $(1/\sqrt{2\pi})$ in both transform and inverse transform (see eqns (2.4) and (2.9)), and in mathematical texts such a convention is common. However in practical applications it often happens that two functions forming a Fourier pair correspond to well-known physical quantities whose definitions are fixed by previous usage, and usually in such cases the unsymmetrical form of eqn (2.14) is forced upon us. Symmetry may also be achieved by introducing a factor in the exponent as in eqns (2.6) and (2.11), but this convention is not very widespread. Choice of a different kind arises with the sign of the exponent, and sometimes the

Fourier transform is given a positive exponent whilst the inverse transform has a negative exponent. In adopting the definitions in eqn (2.14) we are conforming to the usage dominant amongst engineers.

Note when using the arrow, $\leftrightarrow$, that the functions are not interchangeable the one on the right being the transform of its partner, the one on the left being the inverse transform. The nomenclature based on the symbols FT^+ and FT^- is convenient since it allows the Fourier inversion theorem to be written formally as

$$FT^+FT^- = FT^-FT^+ = 1.$$

This is possible because of the following identities which follow straightforwardly from eqn (2.14) above.

$$FT^+\{FT^-\{f(x)\}\} = f(x)$$

$$FT^-\{FT^+\{F(y)\}\} = F(y).$$

Some simple operational algebra is possible. For instance if f and F are two functions (of different variables) such that

$$f = FT^+\{F\}$$

then we may 'multiply through' by FT^- to obtain

$$FT^-\{f\} = F.$$

The variables are not indicated since it is understood that operating with $FT^\pm$ produces a function of the other variable of the pair. In some applications it may not be clear what the pair of variables are, in which case they can be included explicitly in the operator symbol by writing FT^{+xy} or FT^{-xy} as the case may be. We shall use the symbol FT to refer to a transform or inverse transform when the choice is immaterial.

It is sometimes useful to note the following result, proved in Appendix D:

$$FT^-FT^-\{f(x)\} = 2\pi f(-x)$$

$$FT^+FT^+\{f(x)\} = \frac{f(-x)}{2\pi}.$$

The allowed form of $f(x)$. $f(x)$ must be single valued, but may be complex. A full consideration of the further conditions that $f(x)$ must satisfy for the inversion theorem to hold is involved, and the reader is referred to references 12, 15, 20 and 21. It is simpler to state sufficient conditions for $f(x)$ to satisfy rather than necessary conditions. If $F(y)$ is to contain no singularity functions such as delta functions or step discontinuities then it is sufficient that $f(x)$ has (a) only a finite number of maxima or minima, (b) no infinite discontinuities, (c) only a finite number of finite discontinuities, and is absolutely integrable (i.e. $\int_{-\infty}^{+\infty} |f(x)|\, dx < \infty$)†. Note that $f(x)$ may contain a finite number of delta functions.

If singularity functions such as delta functions and finite or infinite discontinuities are allowed in the function $F(y)$ then it is a sufficient condition for the inversion theorem to hold that $f(x)$ may be reduced to the sum of a part satisfying the above conditions plus any or all of (a) a constant term, (b) a finite number of periodic functions, (c) a finite number of step functions. The periodic functions referred to may contain discontinuities and need only satisfy Dirichlet's conditions (p. 3). By a step function we mean simply that $f(x)$ assumes constant but different values either side of some specified value of x. The class of functions allowed is now extremely wide, as may be verified by perusing the table in Section 2.5, certain random functions being the only commonly met functions which do not possess a Fourier transform (see Section 6.2).

Note that a Fourier series expansion of a periodic function can now be regarded as equivalent to a particular kind of Fourier transform, the function $F(y)$ consisting of delta functions at values of y equal to multiples of some fundamental value.

Simplifying conditions. The relationships in eqns (2.7)–(2.13) are simplified when $f(t)$ has special properties, and the following results are easily verified.

If $f(t)$ is real, then: $A(\omega)$, $B(\omega)$, $C(\omega)$, $D(\omega)$, $\phi(\omega)$, $\theta(\omega)$ are real

$E(\omega)$, $F(\omega)$, $G(v)$ are complex

$$F(\omega) = F^*(-\omega) \qquad A(\omega) = \frac{1}{\pi}\,\mathrm{Re}\,F(\omega)$$

$$F^*(\omega) = F(-\omega) \qquad B(\omega) = -\frac{1}{\pi}\,\mathrm{Im}\,F(\omega).$$

If $f(t)$ is pure imaginary then: $A(\omega)$, $B(\omega)$, $C(\omega)$, $D(\omega)$, are pure imaginary

$\phi(\omega)$, $\theta(\omega)$ are real

$E(\omega)$, $F(\omega)$, $G(v)$ are complex

$$F(\omega) = -F^*(-\omega) \qquad A(\omega) = \frac{i}{\pi}\,\mathrm{Im}\,F(\omega)$$

$$F^*(\omega) = -F(-\omega) \qquad B(\omega) = \frac{i}{\pi}\,\mathrm{Re}\,F(\omega).$$

†In cases of ambiguity we mean

$$\lim_{X\to\infty} \int_{-X}^{+X} |f(x)|\, dx.$$

This is known as the Cauchy principal value of the integral.

If $f(t)$ is even then:

$$A(\omega) = C(\omega) = D(\omega) = \frac{1}{\pi} F(\omega) \qquad B(\omega) = 0$$

$$\phi(\omega) = 0 \qquad \theta(\omega) = \pi/2$$

$$F(\omega) = F(-\omega).$$

If $f(t)$ is odd then:

$$B(\omega) = C(\omega) = D(\omega) = \frac{i}{\pi} F(\omega) \qquad A(\omega) = 0$$

$$\phi(\omega) = -\pi/2 \qquad \theta(\omega) = 0$$

$$F(\omega) = -F(-\omega).$$

If $f(t)$ is real and even
       or   } *then*: $F(\omega)$ *is real.*
pure imaginary and odd

If $f(t)$ is real and odd,
       or   } *then*: $F(\omega)$ *is pure imaginary.*
pure imaginary and even

Sine and Cosine transforms. If $f(x)$ only exists for the range $x > 0$, then a special simplification is possible. We may write

$$f(x) = \sqrt{\frac{2}{\pi}} \int_0^\infty F_c(y) \cos xy \, dy \qquad (2.15)$$

$$= \sqrt{\frac{2}{\pi}} \int_0^\infty F_s(y) \sin xy \, dy \qquad (2.16)$$

with

$$F_c(y) = \sqrt{\frac{2}{\pi}} \int_0^\infty f(x) \cos xy \, dx \qquad (2.17)$$

$$F_s(y) = \sqrt{\frac{2}{\pi}} \int_0^\infty f(x) \sin xy \, dx. \qquad (2.18)$$

These forms now possess a complete symmetry, and may be used in situations when x represents time and $f(x)$ represents some force voltage, etc., applied to a system from zero time onwards. The pair of variables x and y then each exist for positive values only, and $F_s(y)$ and $F_c(y)$ are called the sine and cosine transforms respectively of $f(x)$.

The results above may be regarded as the result of constructing from

$f(x)$ an even or odd function by adding appropriate portions to $f(x)$ in the negative x region, and then finding the Fourier transform of the function so constructed. The coefficients $B(\omega)$ and $A(\omega)$ of eqn (2.1) come respectively to be zero for the even and odd cases.

Negative frequencies. It can appear disconcerting in an example such as the acoustic case considered in Section 2.1 to find that the pressure variation $f(t)$ has a transform $F(\omega)$ which requires ω to take on negative values, when clearly in a practical application only positive values of ω have direct meaning. Rather than try to assign any physical significance to the negative frequency it is simpler to note that $F(\omega)$ always appears merely in an intermediate step in a calculation and that directly measurable quantities come to be functions of positive values of ω. For instance if the signal $f(t)$ is being synthesised from an array of oscillators at various frequencies, then the amplitude and phase of the oscillator at frequency ω are given by $(1/\pi)\sqrt{F(\omega)F(-\omega)}$ and by $\tan^{-1}[-i\{F(\omega) - F(-\omega)\}/\{F(\omega) + F(-\omega)\}]$, ω in both cases being considered positive (see eqns (2.12), (2.13)).

There are of course other situations, such as diffraction, in which negative values of both the variables involved have direct physical meaning.

Physical Quantities as the real part of complex quantities. Returning to the acoustic example just discussed it can equally be disconcerting that in a problem involving essentially real variables, the transform comes to be complex. Once again $F(\omega)$ should be treated as a tool in the calculation, the final quantity being real when required. For instance if $f(t)$ is real both the above mentioned functions, $F(\omega)F(-\omega)$ and $-i\{F(\omega) - F(-\omega)\}/\{F(\omega) + F(-\omega)\}$, are real.

Confusion of quite a different kind can arise when physical quantities are represented by the real part of a complex quantity, for instance a harmonic variation with time by the real part of $e^{i\omega t}$, or a damped oscillation by the real part of $e^{i\omega t}e^{-at}$. Care is needed since the FT of the real function is not simply the real part of the FT of the complex function.

If
$$f(t) = \operatorname{Re} m(t)$$

and
$$f(t) \leftrightarrow F(\omega)$$

and
$$m(t) \leftrightarrow M(\omega)$$

then it is easy to show that

$$F(\omega) = \tfrac{1}{2}(M(\omega) + M^*(-\omega)). \tag{2.19}$$

We consider the topic further when discussing spectra, in Section 4.4.

2.4   Relations Between Fourier Pairs

It is very useful to have a feeling for the relationship between a function
and its transform, and for the effect that various operations on a function
have on the transform. This can be achieved to quite an extent by studying
the following results and seeing how they are exemplified in the table of
transforms, Section 2.5. The results (a) to (i) below can be established fairly
readily by writing out the transforms explicitly as integrals; the results
(j) onwards rely on an understanding of the concept of convolution. Convolution is discussed in Sections 5.1 and 5.4 which are arranged to be self
contained for ready reference.

(a) *Addition and subtraction.*

$$FT^{\pm}\{f(x) + g(x)\} = FT^{\pm}f(x) + FT^{\pm}g(x). \qquad (2.20)$$

(b) *Multiplication by a constant.*

$$FT^{\pm}\{af(x)\} = aFT^{\pm}f(x). \qquad (2.21)$$

(c) *Scaling.*

If $\qquad\qquad\qquad\qquad f(x) \leftrightarrow F(y)$

then $\qquad\qquad f(ax) \leftrightarrow \dfrac{1}{|a|} F\left(\dfrac{y}{a}\right) \qquad (a = \text{real constant}) \qquad (2.22)$

and $\qquad\qquad\qquad \dfrac{1}{|a|} f\left(\dfrac{x}{a}\right) \leftrightarrow F(ay). \qquad (2.23)$

When a delta function occurs it is important to remember that $\delta[(y/a) - y_0]$
equals $a\delta(y - ay_0)$ and not simply $\delta(y - ay_0)$, see eqn B.10.

(d) *Shifting.* If a function of x is shifted in the positive direction by an
amount x_0 then the $FT^{\pm}$ is multiplied by $e^{\pm ix_0 y}$. Conversely multiplying a
function of x by $e^{iy_0 x}$ shifts its $FT^{\pm}$ by an amount $\mp y_0$.

If $\qquad\qquad\qquad\qquad f(x) \leftrightarrow F(y)$

then $\qquad\qquad\qquad f(x \pm x_0) \leftrightarrow e^{\pm ix_0 y} F(y) \qquad (2.24)$

and $\qquad\qquad\qquad e^{\pm iy_0 x} f(x) \leftrightarrow F(y \mp y_0). \qquad (2.25)$

(e) *Areas.* The area under a function is equal to the value of its FT^- at the origin, and conversely the value of a function at the origin is equal to $1/2\pi$ times the area of its FT^{-}.

$$\int_{-\infty}^{+\infty} f(x)\,dx = F(0) \qquad f(0) = \frac{1}{2\pi}\int_{-\infty}^{+\infty} F(y)\,dy. \tag{2.26}$$

The areas are thus not equal. However the areas under the squares of the modulae are related simply as follows:

$$\int_{-\infty}^{+\infty} |f(x)|^2\,dx = \frac{1}{2\pi}\int_{-\infty}^{+\infty} |F(y)|^2\,dy. \tag{2.27}$$

This is a special case of Parseval's theorem, proved in Appendix E.

(f) *Widths*. If a function and its transform have the form of a 'hump' centred on the origin then the width of the one hump is inversely proportional to the width of the other hump. If the widths, w and W, are defined in terms of the widths of rectangular functions having the same areas as the functions themselves, as in Fig. 2.1, then it follows from the first results quoted in (e) above that $wW = 2\pi$.

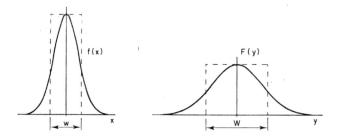

FIGURE 2.1

Whilst the relationship is true for any pair of functions, the interpretation is obviously most useful for positive real functions symmetrically peaked about the origin.

A rather more widely used relation is based on widths defined as the second moments of $|f(x)^2$ and $|F(y)|^2$ about any convenient points x_0 and y_0. The result, proved in Appendix F, is

$$\Delta x \Delta y \geqslant \tfrac{1}{2} \tag{2.28}$$

where

$$(\Delta x)^2 = \int_{-\infty}^{+\infty} (x - x_0)^2 |f(x)|^2 \, dx \bigg/ \int_{-\infty}^{+\infty} |f(x)|^2 \, dx$$

$$(\Delta y)^2 = \int_{-\infty}^{+\infty} (y - y_0)^2 |F(y)|^2 \, dy \bigg/ \int_{-\infty}^{+\infty} |F(y)|^2 \, dy$$

and where x_0 and y_0 are arbitrary.

This result is the mathematical basis for the uncertainty principle in Quantum Mechanics, when the probability amplitudes for position and momentum are related through a Fourier transform. It also relates bandwidth with signal duration in the analysis of oscillatory phenomena. The result is inapplicable to functions which tend to zero too slowly at infinity, such as the Lorentz function (eqns (2.39)). For a further result relating to width see (i) below.

(g) *Symmetry.* If $f(x)$ is an even function, then so is $F(y)$. If $f(x)$ is an odd function, so also is $F(y)$.

(h) *Differentiation.* If a function is differentiated n times then its $FT^{\pm}$ must be multiplied by $(\mp iy)^n$, and conversely if a function of x is multiplied by $(\mp ix)^n$ then its $FT^{\pm}$ must be differentiated n times.

If

$$f(x) \leftrightarrow F(y)$$

then

$$\frac{d^n f(x)}{dx^n} \leftrightarrow (iy)^n F(y) \tag{2.29}$$

and

$$(-ix)^n f(x) \leftrightarrow \frac{d^n F(y)}{dx^n}. \tag{2.30}$$

We assume of course that all functions considered are 'allowed'. One need not be over cautious, since for instance one can validly treat a delta function as resulting from differentiating a step function. This may be verified by taking each as the limiting form of some non-singular function. One may likewise deal with integration by appropriately dividing the $FT^{\pm}$ by $(\mp iy)$; however a delta function at the origin must be inserted whose strength is dependent upon the constant of integration.

(i) *Moments.* The value of the nth differential of a function at the origin is equal to $i^n/2\pi$ times the nth moment of the FT^-. Conversely the nth moment of a function is equal to i^n times the nth differential of its FT^- at the origin.

Bearing in mind that the nth moment of a function $f(x)$ is defined as $\int_{-\infty}^{+\infty} x^n f(x)\,\mathrm{d}x$, we may express the above results by

$$\left[\frac{\mathrm{d}^n f(x)}{\mathrm{d}x^n}\right]_{x=0} = \frac{i^n}{2\pi}\int_{-\infty}^{+\infty} y^n F(y)\,\mathrm{d}y \tag{2.31}$$

$$\int_{-\infty}^{+\infty} x^n f(x)\,\mathrm{d}x = i^n\left[\frac{\mathrm{d}^n F(y)}{\mathrm{d}y^n}\right]_{y=0}. \tag{2.32}$$

The result follows as a combination of eqns (2.26), (2.29) and (2.30) above relating to areas and to differentiation. Note that the zeroth moment of a real function $f(x)$ is its area, the first moment is equal to the product of the area times the distance of the 'centre of gravity' of $f(x)$ from the origin, and the second moment is equal to the 'moment of inertia' of $f(x)$ about the origin.

The result for the second moment means that for a real function consisting of a 'hump' centred on the origin, its mean square width is determined by the value of the second differential of its transform at the origin.

Using eqn (2.32) and a Maclaurin series expansion for the value of a function near the origin, we may obtain the following interesting series which is sometimes useful in evaluating a $FT^{\pm}$ near its origin.

$$F(y) = \int_{-\infty}^{+\infty} f(x)\,\mathrm{d}x \left\{1 - iy\langle x\rangle - \frac{y^2}{2!}\langle x^2\rangle + \frac{iy^3}{3!}\langle x^3\rangle\right.$$

$$\left. + \frac{y^4}{4!}\langle x^4\rangle + \ldots + \frac{(-i)^n y^n}{n!}\langle x^n\rangle + \ldots\right\} \tag{2.33}$$

where

$$\langle x^n\rangle = \int_{-\infty}^{+\infty} x^n f(x)\,\mathrm{d}x \bigg/ \int_{-\infty}^{+\infty} f(x)\,\mathrm{d}x.$$

(j) *Products and convolution.* The FT^- of the product of two functions is proportional to the convolution of their FT^-s. Conversely the FT^- of the convolution of two functions is equal to the product of the individual FT^-s. For a self contained description of the meaning of convolution, and of this particular result, see Sections 5.1 and 5.4. The proof appears in Appendix G.

For reference sake we quote the results in advance.

If $f(x) \leftrightarrow F(y)$ and $g(x) \leftrightarrow G(y)$

then $f(x)g(x) \leftrightarrow \dfrac{1}{(2\pi)}\{F(y) \otimes G(y)\}$ \hfill (2.34)

and $\{f(x) \otimes g(x)\} \leftrightarrow F(y)G(y)$ \hfill (2.35) where the symbol $\otimes$ placed between two functions defines the following operation known as convolution:

$$\{f(x) \otimes g(x)\} = \int_{-\infty}^{+\infty} f(x_1)g(x - x_1)\,dx_1.$$

This most important result is a great help in building up the transform of complicated functions from a knowledge of the transforms of certain basic functions. It is also related directly to important phenomena in diffraction and modulation theory.

(k) *Repetition and sampling.* Repetition of a function *ad infinitum* at regular intervals x_0 removes all of the transform except delta function samples at $y = \pm n(2\pi/x_0)(n = 0, \pm 1, \pm 2, \ldots)$. If the function itself is wider than x_0 then the overlaps must be taken as adding together. Conversely, 'slicing' a function so as to leave only delta function samples at $x = nx_0(n = 0, \pm 1, \pm 2, \ldots)$ has the effect of making the transform repeat itself at intervals $2\pi/x_0$ (with overlaps adding).

Thus if

$$f(x) \leftrightarrow F(y)$$

then

$$\sum_{n=-\infty}^{+\infty} f(x - nx_0) \leftrightarrow \frac{2\pi}{x_0} \sum_{n=-\infty}^{+\infty} \delta\left(y - n\frac{2\pi}{x_0}\right) F(y) \qquad (2.36)$$

and

$$\sum_{n=-\infty}^{+\infty} \delta(x - nx_0)f(x) \leftrightarrow \frac{1}{x_0} \sum_{n=-\infty}^{+\infty} F\left(y - n\frac{2\pi}{x_0}\right). \qquad (2.37)$$

Note that if the chopping is done sufficiently finely, the $FT^{\pm}$ is unaltered except for the repetitive addition of replicas which may be made so far removed that overlap is negligible. Thus for many purposes a knowledge of the values of a function at a regular set of points is as good as a complete knowledge of the function.

The derivation of these results is seen simply if the process of repetition is regarded as the convolution of a function with an infinite set of regularly spaced delta functions. Similarly, 'slicing' results from multiplication by such a set. (See Chapter 5.)

(l) *Truncation and smoothing.* Removing all portions of a function for $|x| > x_0$ has the effect of smoothing structure in the transform finer than $\Delta y \approx 1/x_0$. Similarly smoothing a function to remove structure finer than Δx, will have the effect of making the transform tend to zero for $|y| \gtrsim 1/\Delta x$. These results are a special case of (j) in that truncation means multiplying the given function by some truncating function such as a square function. The result is that the transform becomes convoluted with the transform of the truncating function, this convolution having the effect of smoothing or blurring (see Chapter 5).

2.5 Table of Fourier Transforms

This table is intended to cover most of the functions commonly met in elementary applications and also to exemplify the various results described in Section 2.4. For more comprehensive tabulations, see references 2 and 7. The sketches show the real and/or imaginary parts of the functions, the real parts being drawn on the left. Delta functions are drawn as thick vertical lines, the height representing the 'area' under the delta function. Dashed lines represent imaginary delta functions. Nos 2.38–2.57 do not involve singularity functions and can be obtained using standard integrals. Nos 2.58 onwards involve delta functions and may be based on limiting forms of Nos 2.38–2.57.

In using the table it is useful to note that:

$$FT^{-}\{F(y)\} = 2\pi f(-x) \text{ and } FT^{+}\{f(x)\} = \frac{F(-y)}{2\pi}.$$

Many of the transforms given in the table may in fact be derived from simpler ones using the theorems on shifting, multiplication and convolution. It is a very good exercise to see the different ways in which various transforms may be related, and we list below just one set of such relations. Any function which is periodic may of course be dealt with using the Fourier series techniques.

2.38–2.48 Use standard integrals.

2.49–2.50 Derive from 2.48 by shifting and adding.

2.51–2.54 Derive from 2.48 and 2.58–2.60 using product theorem.

2.55   Derive from 2.48 by convoluting two rectangular functions to give a triangle.

2.56   Use standard integral.

2.57   Derive from 2.48 and 2.55 by subtraction.

2.58–2.60 Limiting cases of 2.42–2.44 ($a \to 0$): or perform inverse transform using delta function integration, eqn B.9.

2.61–2.62 Derive from 2.59–2.60 using product theorem.

2.63 Derive from 2.54 using repetition theorem.

2.64 Derive from 2.63 using shift theorem.

2.65–2.67 Derive from 2.58–2.60 using product theorem.

2.68 Derive from 2.38 ($a \to \infty$).

2.69–2.70 Derive from 2.68 using shift and addition theorems.

2.71 Derive from 2.69 using addition theorem. Note

$$\sum_{n=0}^{N-1} e^{in\alpha} = \left\{ \frac{\sin(N\alpha/2)}{\sin(\alpha/2)} \right\} e^{i(N-1)\alpha/2}$$

2.72–2.73 Derive from 2.71 ($N \to \infty$).

2.74 Derive from 2.72 and 2.59–2.60 using product theorem.

2.75 Derive from 2.38 ($a \to 0$).

2.76 Derive from 2.41 ($a \to 0$).

2.77 Derive from 2.76 and 2.75 by addition.

2.78 Derive from 2.77 and 2.40 by subtraction.

2.79 Derive from 2.75 and 2.48 by subtraction.

2.80–2.87 Use standard integrals (Appendix H).

2.88 Derive from 2.38 ($a^2 \to ia^2$).

2.89 Special case of result proved in Appendix R.

2.90–2.91 Derive from 2.72 using product and convolution theorems.

$$F(y) = \int_{-\infty}^{+\infty} f(x)\, e^{-ixy}\, dx = FT^-\{f(x)\}$$

$F(y)$

$$f(x) = (1/2\pi)\int_{-\infty}^{+\infty} F(y)\, e^{+ixy}\, dy = FT^+\{F(y)\}$$

$f(x)$

$A\exp(-a^2x^2)$ [Gaussian]

$$\frac{A\sqrt{\pi}}{a}\exp(-y^2/4a^2) \qquad (2.38)$$

[Gaussian]

$A\exp(-a|x|)$

$$\frac{2A}{a}\frac{a^2}{a^2+y^2} \qquad (2.39)$$

[Lorentzian]

$$A\exp(-ax) \quad [x>0]$$
$$0 \quad [x<0]$$

$$A\left\{\frac{a-iy}{a^2+y^2}\right\} \qquad (2.40)$$

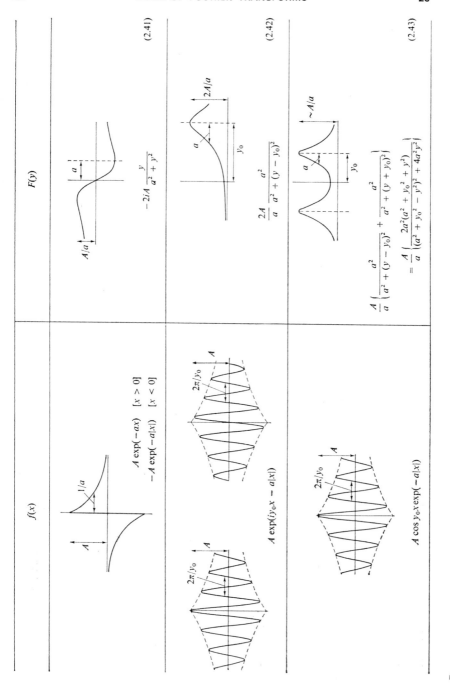

$f(x)$

$F(y)$

$A \exp(-ax) \quad [x > 0]$
$-A \exp(-a|x|) \quad [x < 0]$

$$-2iA \frac{y}{a^2 + y^2} \tag{2.41}$$

$A \exp(iy_0 x - a|x|)$

$$\frac{2A}{a} \frac{a^2}{a^2 + (y - y_0)^2} \tag{2.42}$$

$A \cos y_0 x \exp(-a|x|)$

$$\frac{A}{a} \left\{ \frac{a^2}{a^2 + (y - y_0)^2} + \frac{a^2}{a^2 + (y + y_0)^2} \right\}$$

$$= \frac{A}{a} \left\{ \frac{2a^2(a^2 + y_0^2 + y^2)}{(a^2 + y_0^2 - y^2)^2 + 4a^2 y^2} \right\} \tag{2.43}$$

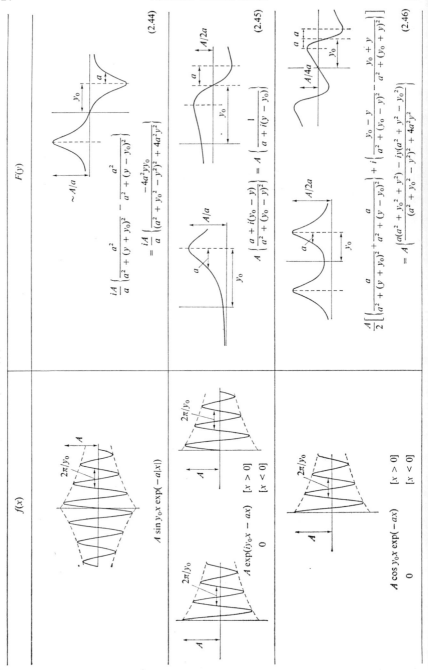

$F(y)$

$$\frac{iA}{a}\left\{\frac{a^2}{a^2+(y+y_0)^2}-\frac{a^2}{a^2+(y-y_0)^2}\right\}$$
$$=\frac{iA}{a}\left\{\frac{-4a^2 y_0 y}{(a^2+y_0^2-y^2)^2+4a^2 y^2}\right\} \tag{2.44}$$

$$A\left\{\frac{a+i(y_0-y)}{a^2+(y_0-y)^2}\right\}=A\left\{\frac{1}{a+i(y-y_0)}\right\} \tag{2.45}$$

$$\frac{A}{2}\left[\left\{\frac{a}{a^2+(y+y_0)^2}+\frac{a}{a^2+(y-y_0)^2}\right\}+i\left\{\frac{y_0-y}{a^2+(y_0-y)^2}-\frac{y_0+y}{a^2+(y_0+y)^2}\right\}\right]$$
$$=A\left\{\frac{a(a^2+y_0^2+y^2)-iy(a^2+y^2-y_0^2)}{(a^2+y_0^2-y^2)^2+4a^2 y^2}\right\} \tag{2.46}$$

$f(x)$

$A\sin y_0 x\exp(-a|x|)$

$A\exp(iy_0 x-ax)$ $[x>0]$
0 $[x<0]$

$A\cos y_0 x\exp(-ax)$ $[x>0]$
0 $[x<0]$

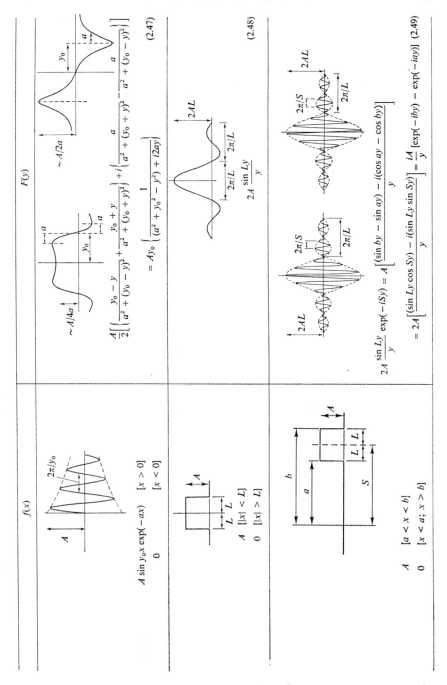

$f(x)$ $F(y)$

$A \sin y_0 x \exp(-ax)$ $[x > 0]$

0 $[x < 0]$

$$\frac{A}{2}\left[\left(\frac{y_0 - y}{a^2 + (y_0 - y)^2} + \frac{y_0 + y}{a^2 + (y_0 + y)^2}\right) + i\left(\frac{a}{a^2 + (y_0 + y)^2} - \frac{a}{a^2 + (y_0 - y)^2}\right)\right] \quad (2.47)$$

A $[|x| < L]$

0 $[|x| > L]$

$$= Ay_0\left\{\frac{1}{(a^2 + y_0^2 - y^2) + i2ay}\right\}$$

$$2A\frac{\sin Ly}{y} \qquad (2.48)$$

A $[a < x < b]$

0 $[x < a; x > b]$

$$2A\frac{\sin Ly}{y}\exp(-iSy) = A\left[\frac{(\sin by - \sin ay) - i(\cos ay - \cos by)}{y}\right]$$

$$= 2A\left[\frac{(\sin Ly \cos Sy) - i(\sin Ly \sin Sy)}{y}\right] = \frac{iA}{y}\left[\exp(-iby) - \exp(-iay)\right] \quad (2.49)$$

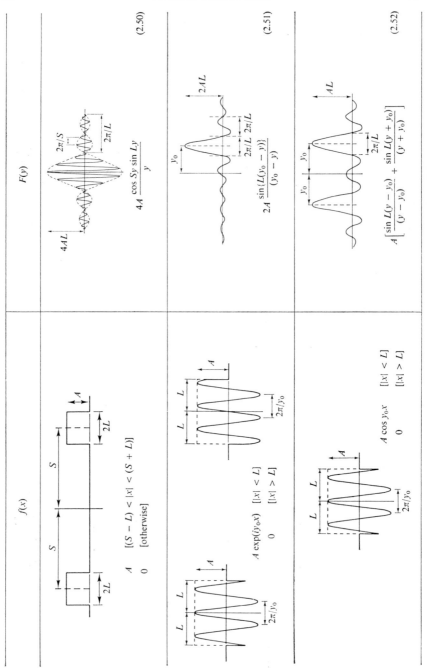

$$f(x)$$

$$A \quad [(S - L) < |x| < (S + L)]$$
$$0 \quad [\text{otherwise}]$$

$$A \exp(iy_0 x) \quad [|x| < L]$$
$$0 \quad [|x| > L]$$

$$A \cos y_0 x \quad [|x| < L]$$
$$0 \quad [|x| > L]$$

$$F(y)$$

$$4A \frac{\cos Sy \sin Ly}{y} \quad (2.50)$$

$$2A \frac{\sin\{L(y_0 - y)\}}{(y_0 - y)} \quad (2.51)$$

$$A\left[\frac{\sin L(y - y_0)}{(y - y_0)} + \frac{\sin L(y + y_0)}{(y + y_0)}\right] \quad (2.52)$$

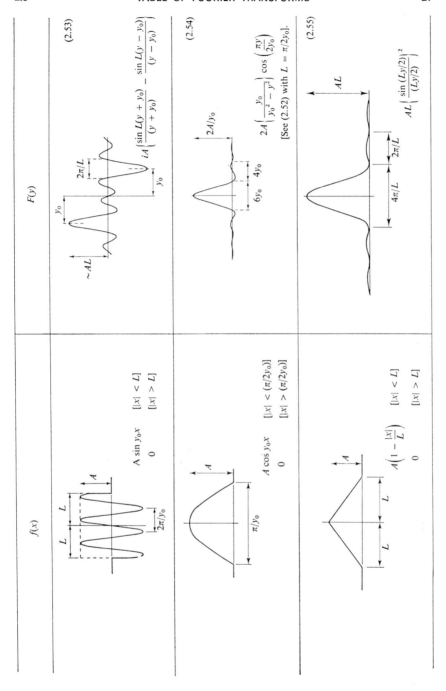

$$f(x)$$

$$A \sin y_0 x \quad [|x| < L]$$
$$0 \quad\quad [|x| > L]$$

$$A \cos y_0 x \quad [|x| < (\pi/2y_0)]$$
$$0 \quad\quad\quad [|x| > (\pi/2y_0)]$$

$$A\left(1 - \frac{|x|}{L}\right) \quad [|x| < L]$$
$$0 \quad\quad\quad\quad [|x| > L]$$

$$F(y)$$

$$iA\left\{\frac{\sin L(y + y_0)}{(y + y_0)} - \frac{\sin L(y - y_0)}{(y - y_0)}\right\} \tag{2.53}$$

$$2A\left(\frac{y_0}{y_0{}^2 - y^2}\right)\cos\left(\frac{\pi y}{2y_0}\right) \tag{2.54}$$
[See (2.52) with $L = \pi/2y_0$].

$$AL\left\{\frac{\sin (Ly/2)}{(Ly/2)}\right\}^2 \tag{2.55}$$

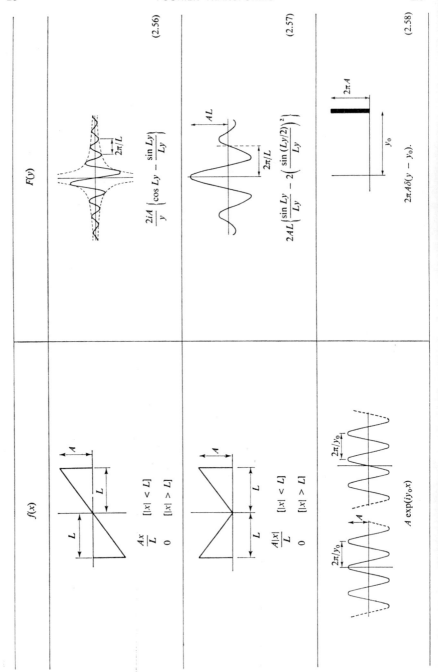

$$F(y)$$

$$\frac{2iA}{y}\left(\cos Ly - \frac{\sin Ly}{Ly}\right) \qquad (2.56)$$

$$2AL\left\{\frac{\sin Ly}{Ly} - 2\left(\frac{\sin (Ly/2)}{Ly}\right)^2\right\} \qquad (2.57)$$

$$2\pi A\delta(y - y_0). \qquad (2.58)$$

$$f(x)$$

$$\frac{Ax}{L} \quad [|x| < L]$$
$$0 \quad [|x| > L]$$

$$\frac{A|x|}{L} \quad [|x| < L]$$
$$0 \quad [|x| > L]$$

$$A \exp(iy_0 x)$$

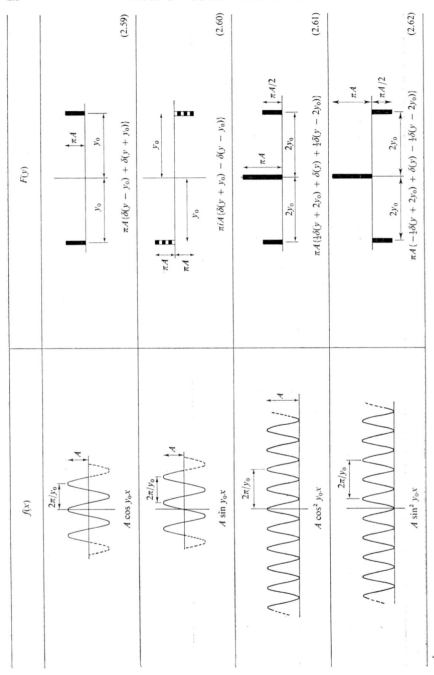

$F(y)$

$\pi A \{\delta(y - y_0) + \delta(y + y_0)\}$ (2.59)

$\pi i A \{\delta(y + y_0) - \delta(y - y_0)\}$ (2.60)

$\pi A \{\tfrac{1}{2}\delta(y + 2y_0) + \delta(y) + \tfrac{1}{2}\delta(y - 2y_0)\}$ (2.61)

$\pi A \{-\tfrac{1}{4}\delta(y + 2y_0) + \delta(y) - \tfrac{1}{4}\delta(y - 2y_0)\}$ (2.62)

$f(x)$

$A \cos y_0 x$

$A \sin y_0 x$

$A \cos^2 y_0 x$

$A \sin^2 y_0 x$

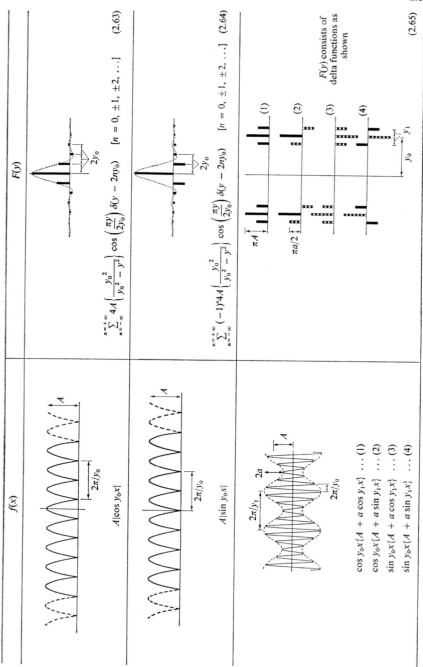

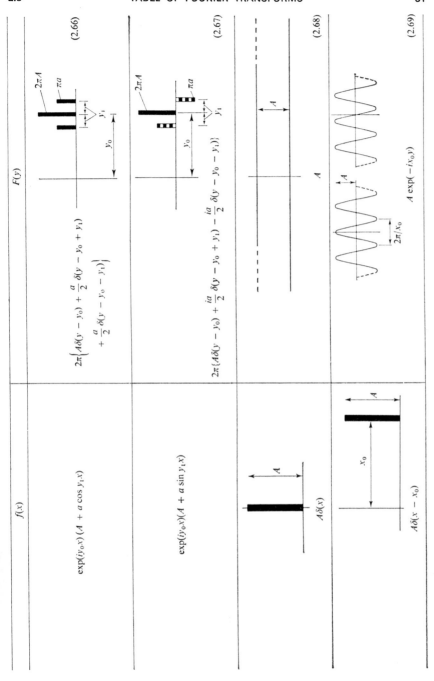

$f(x)$ $F(y)$

$$\exp(iy_0 x)(A + a \cos y_1 x)$$

$$2\pi\left[A\delta(y - y_0) + \frac{a}{2}\delta(y - y_0 + y_1) + \frac{a}{2}\delta(y - y_0 - y_1)\right] \quad (2.66)$$

$$\exp(iy_0 x)(A + a \sin y_1 x)$$

$$2\pi\left\{A\delta(y - y_0) + \frac{ia}{2}\delta(y - y_0 + y_1) - \frac{ia}{2}\delta(y - y_0 - y_1)\right\} \quad (2.67)$$

$$A\delta(x) \qquad A \quad (2.68)$$

$$A\delta(x - x_0) \qquad A\exp(-ix_0 y) \quad (2.69)$$

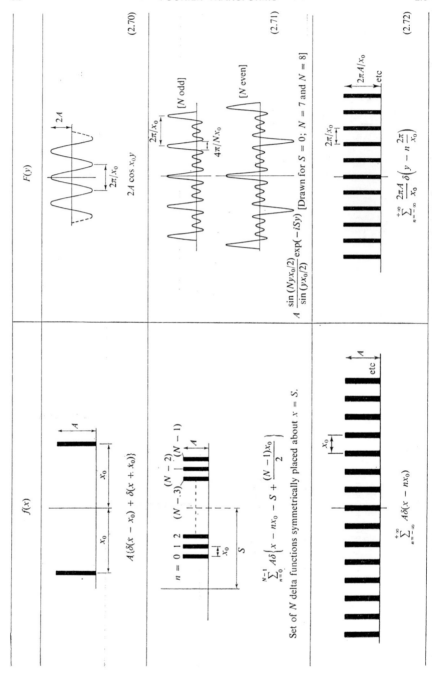

$$F(y)$$

$$2A \cos x_0 y \tag{2.70}$$

$$A \frac{\sin (Nyx_0/2)}{\sin (yx_0/2)} \exp(-iSy) \quad \text{[Drawn for } S = 0; \ N = 7 \text{ and } N = 8\text{]} \tag{2.71}$$

$$\sum_{n=-\infty}^{+\infty} \frac{2\pi A}{x_0} \delta\left(y - n\frac{2\pi}{x_0}\right) \tag{2.72}$$

$$f(x)$$

$$A\{\delta(x - x_0) + \delta(x + x_0)\}$$

$$\sum_{n=0}^{N-1} A\delta\left(x - nx_0 - S + \frac{(N-1)x_0}{2}\right)$$

Set of N delta functions symmetrically placed about $x = S$.

$$\sum_{n=-\infty}^{+\infty} A\delta(x - nx_0)$$

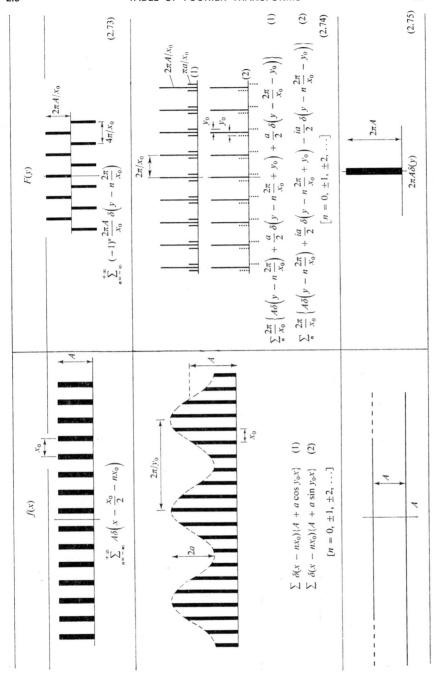

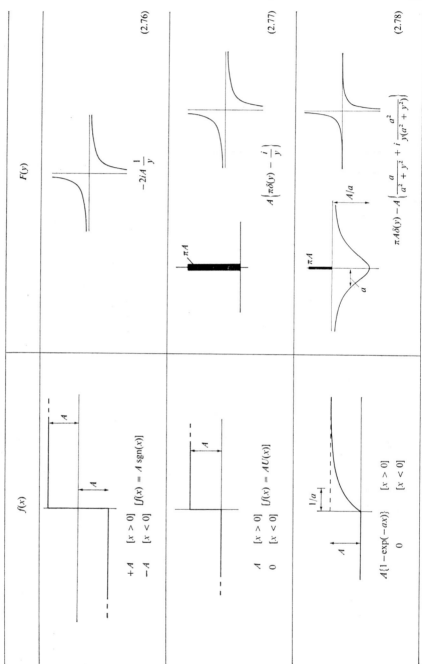

$f(x)$	$F(y)$	
$\begin{array}{ll} A & [\lvert x \rvert > L] \\ 0 & [\lvert x \rvert < L] \end{array}$	$2\pi A\delta(y) - 2A\,\dfrac{\sin Ly}{y}$	(2.79)
$A\exp\{i(a\cos y_0 x + bx)\}$	$2\pi A \displaystyle\sum_{n=-\infty}^{+\infty} (i)^n J_n(a)\delta(y - b - ny_0)$	(2.80)
$A\exp\{i(a\sin y_0 x + bx)\}$	$2\pi A \displaystyle\sum_{n=-\infty}^{+\infty} J_n(a)\delta(y - b - ny_0)\}$	(2.81)

Note: $J_n(-a) = J_{-n}(a) = (-1)^n J_n(a)$. See Appendix H for some properties of Bessel functions.

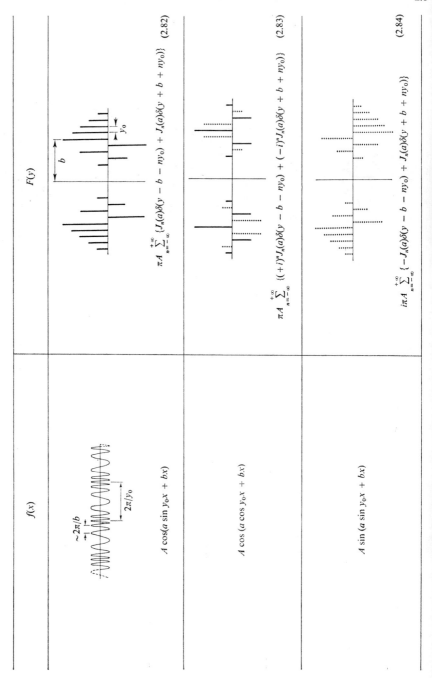

$f(x)$	$F(y)$
$\sim 2\pi/b$ $2\pi/y_0$	b y_0
$A\cos(a\sin y_0 x + bx)$	$\pi A \sum\limits_{n=-\infty}^{+\infty} \{J_n(a)\delta(y - b - ny_0) + J_n(a)\delta(y + b + ny_0)\}$ (2.82)
$A\cos(a\cos y_0 x + bx)$	$\pi A \sum\limits_{n=-\infty}^{+\infty} \{(+i)^n J_n(a)\delta(y - b - ny_0) + (-i)^n J_n(a)\delta(y + b + ny_0)\}$ (2.83)
$A\sin(a\sin y_0 x + bx)$	$i\pi A \sum\limits_{n=-\infty}^{+\infty} \{-J_n(a)\delta(y - b - ny_0) + J_n(a)\delta(y + b + ny_0)\}$ (2.84)

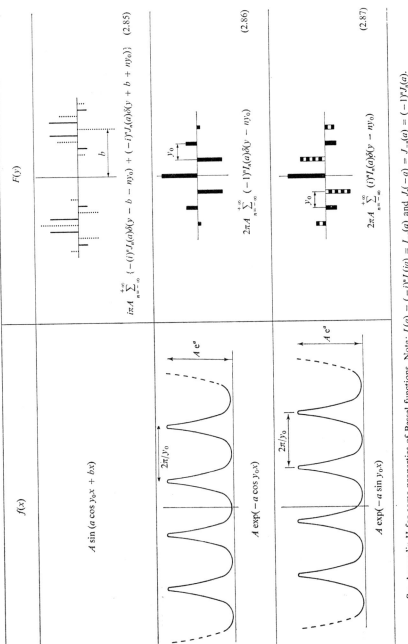

$f(x)$ $F(y)$

$A \sin (a \cos y_0 x + bx)$

$$i\pi A \sum_{n=-\infty}^{+\infty} \{-(i)^n J_n(a)\delta(y - b - ny_0) + (-i)^n J_n(a)\delta(y + b + ny_0)\} \quad (2.85)$$

$A \exp(-a \cos y_0 x)$

$$2\pi A \sum_{n=-\infty}^{+\infty} (-1)^n I_n(a)\delta(y - ny_0) \quad (2.86)$$

$A \exp(-a \sin y_0 x)$

$$2\pi A \sum_{n=-\infty}^{+\infty} (i)^n I_n(a)\delta(y - ny_0) \quad (2.87)$$

See Appendix H for some properties of Bessel functions. Note: $I_n(a) = (-i)^n J_n(ia) = (-i)^n J_n(ia)$ and $J_n(-a) = I_{-n}(a) = J_{-n}(a) = (-1)^n J_n(a)$.

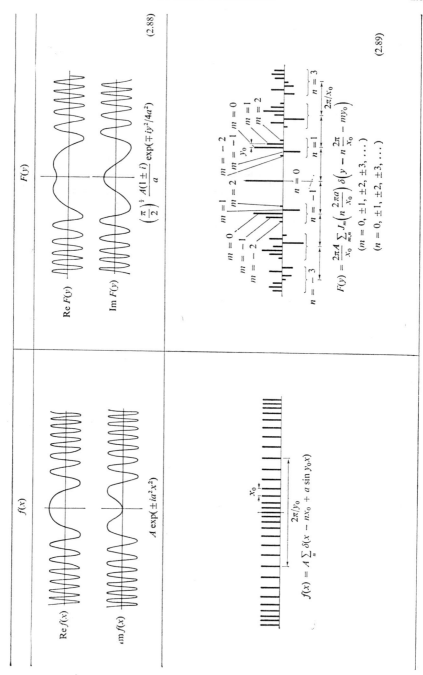

$f(x)$ $F(y)$

$\mathrm{Re}\, f(x)$ $\mathrm{Im}\, f(x)$

$$A \exp(\pm ia^2 x^2)$$

$\mathrm{Re}\, F(y)$ $\mathrm{Im}\, F(y)$

$$\left(\frac{\pi}{2}\right)^{\frac{1}{2}} \frac{A(1 \pm i)}{a} \exp(\mp iy^2/4a^2) \tag{2.88}$$

$$f(x) = A \sum_n \delta(x - nx_0 + a \sin y_0 x)$$

$$F(y) = \frac{2\pi A}{x_0} \sum_{m,n} J_m\left(n\frac{2\pi a}{x_0}\right) \delta\left(y - n\frac{2\pi}{x_0} - my_0\right)$$
$$(m = 0, \pm 1, \pm 2, \pm 3, \ldots)$$
$$(n = 0, \pm 1, \pm 2, \pm 3, \ldots) \tag{2.89}$$

$$f(x) = h(x) \sum_{n=-\infty}^{+\infty} g(x - nx_0)$$

$$f(x) = \sum_{n=-\infty}^{+\infty} h(nx_0)g(x - nx_0)$$

$$F(y) = \frac{1}{x_0} \sum_{n=-\infty}^{+\infty} \left\{ G\left(\frac{n2\pi}{x_0}\right) H\left(y - \frac{n2\pi}{x_0}\right) \right\} \qquad (2.90)$$

$$F(y) = \frac{1}{x_0} G(y) \sum_{n=-\infty}^{+\infty} H\left(y - \frac{n2\pi}{x_0}\right) \qquad (2.91)$$

Chapter 3

Fourier Integrals in Several Dimensions

3.1 The Basic Result in n Dimensions

If

$$
\left.
\begin{aligned}
&f(x_1, x_2 \ldots x_n) \\
&\quad = \frac{1}{(2\pi)^n} \iint \ldots \int F(y_1, y_2 \ldots y_n)\, e^{+i(x_1 y_1 + x_2 y_2 + \ldots x_n y_n)}\, \mathrm{d}y_1 \ldots \mathrm{d}y_n \\
&\text{then} \\
&F(y_1, y_2 \ldots y_n) \\
&\quad = \iint \ldots \int f(x_1, x_2 \ldots x_n)\, e^{-i(x_1 y_1 + x_2 y_2 + \ldots x_n y_n)}\, \mathrm{d}x_1 \ldots \mathrm{d}x_n
\end{aligned}
\right\} \quad (3.1)
$$

This follows as a natural extension of the one-dimensional result eqn (2.14) and is readily verified if the above expressions are broken down into one-dimensional transforms as follows:

$$
f(x_1 \ldots x_n) = FT^{+x_1 y_1}\, FT^{+x_2 y_2} \ldots FT^{+x_n y_n}\{F(y_1 \ldots y_n)\}
$$

$$
F(y_1 \ldots y_n) = FT^{-x_1 y_1}\, FT^{-x_2 y_2} \ldots FT^{-x_n y_n}\{f(x_1 \ldots x_n)\}
$$

where for instance $FT^{-x_1 y_1}$ and $FT^{+x_1 y_1}$ represent one-dimensional transforms (forward and inverse respectively) with x_1 and y_1 as conjugate variables. With $f(x_1, x_2 \ldots x_n)$ and $F(y_1, y_2 \ldots y_n)$ related as in eqn (3.1) we will write them as a Fourier pair as follows

$$
f(x_1 \ldots x_n) \leftrightarrow F(y_1 \ldots y_n). \tag{3.2}
$$

Other conventions are possible with the signs in the exponents changed, with mixed signs in each exponent, or with a factor $(2\pi)^{-n/2}$ placed before the integral sign in both forward and inverse transforms giving symmetry in this respect.

3.2 Use of Vectors

Often in physical applications a three-dimensional transform is used in which x_1, x_2, x_3 are the cartesian components of one vector, and y_1, y_2, y_3 are those of another vector. Letting $\mathbf{r}$ be one vector, with components x, y, z and $\mathbf{k}$ be the other vector, with components k_x, k_y, k_z, we can use the notation:

$$F(\mathbf{k}) = FT^{-\mathbf{k}\cdot\mathbf{r}}\{f(\mathbf{r})\} = \iiint f(\mathbf{r})\,e^{-i\mathbf{k}\cdot\mathbf{r}}\,d\mathbf{r} \qquad (3.3)$$

$$f(\mathbf{r}) = FT^{+\mathbf{k}\cdot\mathbf{r}}\{F(\mathbf{k})\} = \frac{1}{(2\pi)^3} \iiint F(\mathbf{k})\,e^{+i\mathbf{k}\cdot\mathbf{r}}\,d\mathbf{k} \qquad (3.4)$$

$$f(\mathbf{r}) \leftrightarrow F(\mathbf{k}). \qquad (3.5)$$

The integrals are over all values of the respective vectors.

With $\mathbf{r}$ representing a position vector, $e^{i\mathbf{k}\cdot\mathbf{r}}$ is the complex representation for a static harmonic plane wave in three dimensions of wave number $\mathbf{k}$ (i.e. of wavelength $\lambda = 2\pi/|\mathbf{k}|$) and with wave fronts perpendicular to $\mathbf{k}$. Equation (3.4) then represents the possibility of expressing a quantity $f(\mathbf{r})$, varying in any way with position, as an integral sum of harmonic plane wave variations, all wavelengths and directions being considered.

The four-dimensional transform

$$F(\mathbf{k}, \omega) = FT^{-\mathbf{k}\cdot\mathbf{r}-\omega t}\{f(\mathbf{r}, t)\} = \iiiint f(\mathbf{r}, t)\,e^{-i(\mathbf{k}\cdot\mathbf{r}+\omega t)}\,d\mathbf{r}\,dt \qquad (3.6)$$

$$f(\mathbf{r}, t) = FT^{+\mathbf{k}\cdot\mathbf{r}+\omega t}\{F(\mathbf{k}, \omega)\} = \frac{1}{(2\pi)^4} \iiiint F(\mathbf{k}, \omega)\,e^{+i(\mathbf{k}\cdot\mathbf{r}+\omega t)}\,d\mathbf{k}\,d\omega \quad (3.7)$$

with $\mathbf{k}$ and $\mathbf{r}$ as above and ω and t representing angular frequency and time, represents the possibility of expressing a time-dependent function of position $f(\mathbf{r}, t)$ as an integral sum of travelling harmonic plane waves, all wavelengths, directions, and frequencies being considered. Since $e^{\pm i(\mathbf{k}\cdot\mathbf{r}+\omega t)}$ represents a wave travelling in the negative $\mathbf{k}$ direction, whilst $e^{\pm i(\mathbf{k}\cdot\mathbf{r}-\omega t)}$ represents a wave travelling in the positive $\mathbf{k}$ direction, there is some case for using a convention with mixed signs in the exponent. Note that this synthesis is not possible if waves of fixed velocity are considered.

3.3 Simple Operational Results

It is easy to verify that multidimensional transforms can be carried out in stages as single-dimensional transforms. For instance (using a three-dimensional case purely as an example)

$$FT^{-\mathbf{k}\cdot\mathbf{r}}\{f(\mathbf{r})\} = FT^{-xk_x}FT^{-yk_y}FT^{-zk_z}\{f(\mathbf{r})\}.$$

This leads naturally to the operational equation:

$$FT^{-\mathbf{k}\cdot\mathbf{r}}FT^{+\mathbf{k}\cdot\mathbf{r}} = 1.$$

This is useful in working out multidimensional transforms from known one-dimensional forms, and is also sometimes used with four-dimensional transforms. In this latter case the transform is sometimes carried out in two stages, an intermediate function $I(\mathbf{k}, t)$ being introduced thus:

$$I(\mathbf{k}, t) = FT^{+\omega t}F(\mathbf{k}, \omega)$$

so that

$$f(\mathbf{r}, t) = FT^{+\mathbf{k}\cdot\mathbf{r}}I(\mathbf{k}, t).$$

This result follows trivially from the operational results above.

3.4 Delta Functions in 3 Dimensions

The point delta function $\delta(\mathbf{r})$ is an obvious extension of the one-dimensional form (see Appendix B) having the properties

$$\left.\begin{array}{ll} \delta(\mathbf{r}) = 0 & [\mathbf{r} \neq 0] \\ \delta(\mathbf{r}) = \infty & [\mathbf{r} = 0] \\ \int\delta(\mathbf{r})\,d\mathbf{r} = 1 \end{array}\right\} \text{ imprecise}$$

$$\int f(\mathbf{r})\delta(\mathbf{r})\,d\mathbf{r} = f(0) \qquad \} \text{ rigorous.}$$

The integrals here extend over all values of the vector $\mathbf{r}$ (i.e. over all space if $\mathbf{r}$ is a position vector). $\delta(\mathbf{r} - \mathbf{r}_0)$ is similarly a point delta function at the point $\mathbf{r} = \mathbf{r}_0$.

However more forms of delta function are possible, having no analogue in one dimension. For instance consider a function which equals zero everywhere except on a line or a plane (line or plane delta functions) or on some other defined surface such as a spherical shell. Such a function can be classed as a delta function, and its correct use established by considering it as the limiting case of an ordinary function. We shall refer to the integrated value of such a function over all space (or over some appropriate region of space) as its strength, this being analogous to the area under a one-dimensional delta function.

Whereas at first sight it might seem natural that a delta function integrating over all space to give a strength of unity should be chosen as a unit plane delta function, it in fact turns out to be more convenient to work in terms of strength per unit area (even though the integral over all space of such a function would then be infinite if it had unit strength per unit area).

Similarly a line delta function will be characterised by its strength per unit length.

With surfaces of finite area (such as a spherical shell) it is possible to work with either total strength or strength per unit area, and it must be made clear at the time what meaning is attached to a particular function. Similarly with lines of finite length.

The nomenclature adopted will vary slightly with the context, but will usually be self explanatory. For instance the expression

$$f(\mathbf{r}) = A\delta(\mathbf{r} - \mathbf{R}) \qquad [\mathbf{R} \cdot \mathbf{a} = 0]$$

is taken to mean that $f(\mathbf{r})$ is zero if $\mathbf{r}$ does not lie on the plane through the origin perpendicular to $\mathbf{a}$, and that $f(\mathbf{r})$ equals infinity if $\mathbf{r}$ does lie on this plane, in such a way that the integrated value of $f(\mathbf{r})$ over a finite volume (such as a cube) enclosing any unit area of the plane is equal to A.

If a function $f(\mathbf{r})$ is the product of two functions $g(\mathbf{r})$ and $h(\mathbf{r})$ which are allowed to tend towards line and plane delta functions respectively, then $f(\mathbf{r})$ tends to a point delta function at the intersection. It is not difficult to show that the strength of the resulting point delta function is $AB|\mathbf{a}| \, |\mathbf{b}|/\mathbf{a} \cdot \mathbf{b}$ where A and B are the strengths per unit length and per unit area respectively of $g(\mathbf{r})$ and $h(\mathbf{r})$, $\mathbf{a}$ is any vector parallel to the line delta function, and $\mathbf{b}$ is any vector perpendicular to the plane delta function. Colloquially we may say that the line and the plane intersect to give a point of strength $AB|\mathbf{a}| \, |\mathbf{b}|/\mathbf{a} \cdot \mathbf{b}$.

Similarly, with obvious meanings, it can be shown that three planes can intersect to give a point of strength $ABC|\mathbf{a}| \, |\mathbf{b}| \, |\mathbf{c}|/|\mathbf{a} \cdot \mathbf{b} \times \mathbf{c}|$ where A, B, C are the strengths per unit area, and $\mathbf{a}$, $\mathbf{b}$ and $\mathbf{c}$ are any vectors perpendicular to the three planes.

Similarly also two planes intersect to give a line of strength $AB|\mathbf{a}| \, |\mathbf{b}|/|\mathbf{a} \times \mathbf{b}|$ where A and B are the strengths per unit area, and $\mathbf{a}$ and $\mathbf{b}$ are vectors perpendicular to the planes.

These results may be obtained by considering the planes as limiting cases of functions which equal some constant within a sheet of finite thickness and which equal zero outside, and allowing the thickness of the sheet to tend to zero whilst the value of the constant tends to infinity. The vector products in the above results arise from a geometrical consideration of the volumes of intersection of such sheets.

As three-dimensional transforms arise frequently in scattering problems, these delta functions will often arise in the representation of point scatterers, or the uniform distribution of scattering material over a surface or along a line.

3.5   Delta Functions in Two Dimensions

These arise as point or line functions. Using concepts derived in an obvious way from those in Section 3.4 we have the following result. Two line delta functions intersect to give a point delta function of strength $AB/\sin \alpha$ where A and B are the strengths per unit length, and α is the angle between the line delta functions. This can also be written as $AB|\mathbf{a}|\,|\mathbf{b}|/|\mathbf{a} \times \mathbf{b}|$ where $\mathbf{a}$ and $\mathbf{b}$ are vectors both lying in the plane and being perpendicular respectively to the two lines; the meaning is clear although we are perhaps using the vector notation in a doubtful fashion since the vector $\mathbf{a} \times \mathbf{b}$ cannot exist in a two-dimensional system.

3.6   Relations Between Fourier Pairs in Several Dimensions

Many of the results in Section 2.4 for one-dimensional transforms can be generalised in an obvious fashion. We give a selection of the more useful relations below, together with some further relations which have no counterpart in one dimension.

(a) *Addition and subtraction*

$$FT^{\pm}\{f(\mathbf{r}) + g(\mathbf{r})\} = FT^{\pm}\{f(\mathbf{r})\} + FT^{\pm}\{g(\mathbf{r})\}. \tag{3.8}$$

Similarly for functions of any number of variables.

(b) *Multiplication by a constant*

$$FT^{\pm}\{af(\mathbf{r})\} = aFT^{\pm}\{f(\mathbf{r})\}. \tag{3.9}$$

Similarly for functions of any number of variables.

(c) *Scaling*

If
$$f(\mathbf{r}) \leftrightarrow F(\mathbf{k})$$

then
$$f(\mathbf{r}/a) \leftrightarrow |a|^3 F(a\mathbf{k}) \quad \text{for real '}a\text{'} \tag{3.10}$$

and
$$f(a\mathbf{r}) \leftrightarrow \frac{1}{|a|^3} F(\mathbf{k}/a). \tag{3.11}$$

In n dimensions replace $|a|^3$ by $|a|^n$.

(d) *Separable variables*

If $f(\mathbf{r})$ can be expressed as the product of three functions which depend only on x, y and z respectively, then the transform is the product of the individual transforms.

If
$$f(\mathbf{r}) = l(x)m(y)n(z)$$
$$l(x) \leftrightarrow L(k_x)$$
$$m(y) \leftrightarrow M(k_y)$$
$$n(z) \leftrightarrow N(k_z)$$

then
$$f(\mathbf{r}) \leftrightarrow L(k_x)M(k_y)\,N(k_z). \tag{3.12}$$

In such a case the multidimensional transform can thus be built up from a knowledge of one-dimensional transforms. Analogous results hold in any number of dimensions.

(e) *Shifting*

If a function $f(\mathbf{r})$ is shifted by an amount $\mathbf{r}_0$ then its $FT^{\pm}$ must be multiplied by $e^{\pm i\mathbf{k}\cdot\mathbf{r}_0}$. Conversely multiplying a function $f(\mathbf{r})$ by $e^{i\mathbf{k}_0\cdot\mathbf{r}}$ shifts its $FT^{\pm}$ by an amount $\mp\mathbf{k}_0$.

If
$$f(\mathbf{r}) \leftrightarrow F(\mathbf{k})$$

then
$$f(\mathbf{r} \pm \mathbf{r}_0) \leftrightarrow e^{\pm i\mathbf{k}\cdot\mathbf{r}_0}F(\mathbf{k}) \tag{3.13}$$

and
$$e^{\pm i\mathbf{k}_0\cdot\mathbf{r}}f(\mathbf{r}) \leftrightarrow F(\mathbf{k} \mp \mathbf{k}_0). \tag{3.14}$$

Analogous results hold with other numbers of dimensions.

(f) *Value of transforms at the origin. Integrated value*

The integrated value of a function over all values of its n variables equals the value of the FT^{-} at the origin. Conversely the value of a function at the origin is equal to $1/(2\pi)^n$ times the integrated value of its FT^{-}. The integrated values of the squares of the moduli, however, come out to be proportional to each other. For the case of n dimensions we thus have:

$$\left. \begin{array}{l} \displaystyle\int f(\mathbf{r})\,d\mathbf{r} = F(0). \\[2em] \displaystyle f(0) = \frac{1}{(2\pi)^n}\int F(\mathbf{k})\,d\mathbf{k} \end{array} \right\} \tag{3.15}$$

and
$$\int |f(\mathbf{r})|^2\,d\mathbf{r} = \frac{1}{(2\pi)^n}\int |F(\mathbf{k})|^2\,d\mathbf{k} \tag{3.16}$$

where $\mathbf{r}$ and $\mathbf{k}$ represent n-dimensional vectors. Equation (3.16) is a generalised form of Parseval's theorem (Appendix E).

(g) *Symmetry in many dimensions*

If $\mathbf{r}$ and $\mathbf{k}$ represent n-dimensional vectors, and if $f(\mathbf{r})$ is a function only of the modulus of $\mathbf{r}$, then $F(\mathbf{k})$ is a function only of the modulus of $\mathbf{k}$. Moreover the multidimensional integrals can be reduced to single integrals. In two and three dimensions simple geometrical arguments (see Appendix I) can be used to show that:

in two dimensions

$$\iint e^{\pm i\mathbf{k}\cdot\mathbf{r}} f(r)\,d\mathbf{r} = \int_0^\infty 2\pi r f(r) J_0(kr)\,dr \tag{3.17}$$

in three dimensions

$$\iiint e^{\pm i\mathbf{k}\cdot\mathbf{r}} f(r)\,d\mathbf{r} = \int_0^\infty f(r)\,\frac{\sin kr}{kr}\,4\pi r^2\,dr. \tag{3.18}$$

The first result is basic to the diffraction of a plane wave front by a circularly symmetric aperture, the second is basic to the diffraction of a plane wave front by a spherically symmetric object.

These are special cases of a more general result valid for n-dimensions. Using $f(\mathbf{r})$ and $F(\mathbf{k})$ to represent such symmetrical functions in n-dimensions we have

$$\int e^{\pm i\mathbf{k}\cdot\mathbf{r}} f(r)\,d\mathbf{r} = (2\pi)^{n/2} k^{(1-\frac{1}{2}n)} \int_0^\infty r[r^{(\frac{1}{2}n-1)} f(r)] \, J_{(\frac{1}{2}n-1)}(kr)\,dr. \tag{3.19}$$

We may rephrase this in terms of Hankel transforms by saying that $(2\pi)^{-n/2} k^{(\frac{1}{2}n-1)} F(k)$ is the Hankel transform (of order $\frac{1}{2}n - 1$) of $r^{(\frac{1}{2}n-1)} f(\mathbf{r})$. A Hankel transform pair of functions $g(x)$ and $h(y)$ (of order n) are defined by the result that

if
$$g(x) = \int_0^\infty y\,h(y) J_n(xy)\,dy \left.\vphantom{\int_0^\infty}\right\}$$

$$\tag{3.20}$$

then
$$h(y) = \int_0^\infty x\,g(x) J_n(xy)\,dx. \left.\vphantom{\int_0^\infty}\right.$$

For the simple properties of the Bessel functions $J_n(x)$ see Appendix H. For a discussion of the n-dimensional result and Hankel transforms see reference (15).

(h) *Products and convolutions*

In n-dimensions the FT^- of the product of two functions is equal to $(1/2\pi)^n$ times the convolution of the individual FT^-s. Conversely the FT^- of the convolution of two functions is equal to the product of the individual FT^-s. For the meaning of convolution see Chapter 5 and Appendix G. For reference sake we quote the results:

If $\qquad\qquad f(\mathbf{r}) \leftrightarrow F(\mathbf{k})$ and $g(\mathbf{r}) \leftrightarrow G(\mathbf{k})$

then $\qquad\qquad f(\mathbf{r})g(\mathbf{r}) \leftrightarrow \dfrac{1}{(2\pi)^n} \{F(\mathbf{k}) \otimes G(\mathbf{k})\}$ $\qquad\qquad$ (3.21)

and $\qquad\qquad f(\mathbf{r}) \otimes g(\mathbf{r}) \leftrightarrow F(\mathbf{k})G(\mathbf{k})$ $\qquad\qquad$ (3.22)

where the symbol $\otimes$ meaning convolution is defined by

$$f(\mathbf{r}) \otimes g(\mathbf{r}) = \int f(\mathbf{r}_1)g(\mathbf{r} - \mathbf{r}_1)\, d\mathbf{r}.$$

In the above $\mathbf{r}$ and $\mathbf{k}$ refer to n-dimensional vectors, and the integration is over all values of $\mathbf{r}$.

(i) *Expansion for $F(\mathbf{k})$ in terms of moments*

The expansion given in eqn (2.33) receives its multi-dimensional analogue as follows, obtained by expanding the complex exponential as a series. As previously we use $\mathbf{k}$ and $\mathbf{r}$ to represent vectors in a space of n-dimensions.

$$F(\mathbf{k}) = \int e^{-i\mathbf{k}\cdot\mathbf{r}}f(\mathbf{r})\, d\mathbf{r} = \int \left\{ 1 - i(\mathbf{k}\cdot\mathbf{r}) - \frac{(\mathbf{k}\cdot\mathbf{r})^2}{2!} + \dots \right\} f(\mathbf{r})\, d\mathbf{r}$$

$$= \left\{ 1 - i\langle\mathbf{k}\cdot\mathbf{r}\rangle - \frac{\langle(\mathbf{k}\cdot\mathbf{r})^2\rangle}{2!} + \dots + \frac{(-i)^n}{n!}\langle(\mathbf{k}\cdot\mathbf{r})^n\rangle \right\} \int f(\mathbf{r})\, d\mathbf{r}$$
$$\qquad\qquad (3.23)$$

where

$$\langle(\mathbf{k}\cdot\mathbf{r})^n\rangle = \int (\mathbf{k}\cdot\mathbf{r})^n f(\mathbf{r})\, d\mathbf{r} \bigg/ \int f(\mathbf{r})\, d\mathbf{r}$$

3.7   Table of Two-Dimensional Transforms

In the following list we give a few examples showing how two-dimensional transforms may be built up from simpler one-dimensional transforms using the theorems on separable variables, symmetry, convolution and products. We give in each case a hint showing one of the many possible ways of deriving each result.

With $\mathbf{r}$ and $\mathbf{k}$ representing vectors in two dimensions

$$F(\mathbf{k}) = \iint e^{-i\mathbf{k}\cdot\mathbf{r}} f(\mathbf{r})\, d\mathbf{r}$$

$$f(\mathbf{r}) = \frac{1}{(2\pi)^2} \iint e^{+i\mathbf{k}\cdot\mathbf{r}} F(\mathbf{k})\, d\mathbf{k}. \tag{3.24}$$

Gaussian Function

$$f(\mathbf{r}) = A\, e^{-a^2 r^2}$$

$$F(\mathbf{k}) = \frac{A\pi}{a^2}\, e^{-k^2/4a^2}. \tag{3.25}$$

Derived from eqns (2.38) and (3.12). a^2 may be real or imaginary, so that we have the related pair:

$$f(\mathbf{r}) = A\, e^{\pm ia^2 r^2}$$

$$F(\mathbf{k}) = \pm \frac{iA\pi}{a^2}\, e^{\mp ik^2/4a^2}. \tag{3.25a}$$

Rectangular aperture

$$f(\mathbf{r}) = A \qquad [|x| < a \quad \text{and} \quad |y| < b]$$

$$= 0 \qquad \text{otherwise}$$

$$F(\mathbf{k}) = 4A \left\{ \frac{\sin ak_x}{k_x} \frac{\sin bk_y}{k_y} \right\}. \tag{3.26}$$

Derived from eqns (2.48) and (3.12).

Circular disc, radius R

$$f(\mathbf{r}) = A \qquad [|r| < R]$$

$$= 0 \qquad \text{otherwise}$$

$$F(\mathbf{k}) = 2\pi AR^2 \left\{ \frac{J_1(kR)}{kR} \right\}. \tag{3.27}$$

Derived from eqn (3.17) and Appendix H.

Delta function at $\mathbf{r} = \mathbf{r}_0$

$$f(\mathbf{r}) = A\delta(\mathbf{r} - \mathbf{r}_0)$$

$$F(\mathbf{k}) = A\, e^{-i\mathbf{k}\cdot\mathbf{r}}. \tag{3.28}$$

Derived from eqns (2.69) and (3.12), or verify by direct integration.

Circular line delta function of radius R and strength A

$$f(\mathbf{r}) = \frac{A}{2\pi Ra} \qquad [R < r < (R + a); \quad a \ll R]$$

$$= 0 \qquad \text{otherwise}$$

$$F(\mathbf{k}) = AJ_0(kR).$$

(3.29)

Derived from eqn (3.17).

Line delta function from $\mathbf{r} = -\mathbf{a}$ *to* $\mathbf{r} = +\mathbf{a}$ *of strength A per unit length*

$$f(\mathbf{r}) = A\delta(\mathbf{r} - \mathbf{R}) \qquad [\mathbf{R} = \alpha\mathbf{a}; \ |\alpha| < 1]$$

$$F(\mathbf{k}) = 2A\mathbf{a} \left\{ \frac{\sin \mathbf{k} \cdot \mathbf{a}}{\mathbf{k} \cdot \mathbf{a}} \right\}.$$

(3.30)

Derived from eqn (3.26) $(b \to 0)$.

Row of N delta functions, spacing **a.**

$$f(\mathbf{r}) = \sum_n A\delta(\mathbf{r} - n\mathbf{a}) \qquad [n = 0, \pm 1, \pm 2 \ldots \pm (N - 1)/2 \text{ for } N \text{ odd}$$

$$= \pm\tfrac{1}{2}, \pm\tfrac{3}{2}, \ldots \pm (N - 1)/2 \text{ for } N \text{ even}]$$

$$F(\mathbf{k}) = \frac{A \sin (N\mathbf{k} \cdot \mathbf{a}/2)}{\sin (\mathbf{k} \cdot \mathbf{a}/2)}.$$

(3.31)

Derived from eqns (2.71) and (3.12).

Aperture with sides 2**a** *and* 2**b** *centred on origin*

$$f(\mathbf{r}) = A \qquad [\mathbf{r} = \alpha\mathbf{a} + \beta\mathbf{b}; \ |\alpha| < 1, \ |\beta| < 1]$$

$$= 0 \qquad \text{otherwise}$$

$$F(\mathbf{k}) = 4A|\mathbf{a} \times \mathbf{b}| \left\{ \frac{\sin \mathbf{k} \cdot \mathbf{a}}{\mathbf{k} \cdot \mathbf{a}} \ \frac{\sin \mathbf{k} \cdot \mathbf{b}}{\mathbf{k} \cdot \mathbf{b}} \right\}.$$

(3.32)

Derived from eqn (3.30) by convoluting two lines to form an aperture.

Finite lattice of delta functions, N_1 *by* N_2

$$f(\mathbf{r}) = A \sum_{n_1, n_2} \delta(\mathbf{r} - n_1\mathbf{a} - n_2\mathbf{b})$$

$$[n_1 = 0, \pm 1, \pm 2, \ldots \pm (N_1 - 1)/2 \quad \text{for} \quad N_1 \text{ odd}$$

$$= \pm\tfrac{1}{2}, \pm\tfrac{3}{2}, \ldots \pm (N_1 - 1)/2 \text{ for } N_1 \text{ even. Similarly for } n_2].$$

(3.33)

$$F(\mathbf{k}) = A \left\{ \frac{\sin (N_1\mathbf{k} \cdot \mathbf{a}/2)}{\sin (\mathbf{k} \cdot \mathbf{a}/2)} \ \frac{\sin (N_2\mathbf{k} \cdot \mathbf{b}/2)}{\sin (\mathbf{k} \cdot \mathbf{b}/2)} \right\}.$$

Derived from eqn (3.31) by convoluting two rows of delta functions to form a lattice.

Line delta function in direction **a** *of strength A per unit length*

$$f(\mathbf{r}) = A\delta(\mathbf{r} - \mathbf{R}) \qquad [\mathbf{R} = \alpha\mathbf{a}; \ -\infty < \alpha < +\infty]$$

$$F(\mathbf{k}) = 2\pi A\delta(\mathbf{k} - \mathbf{K}) \qquad [\mathbf{K} = \beta\mathbf{a}^*; \ -\infty < \beta < +\infty] \qquad (3.34)$$

$$\mathbf{a}^* \cdot \mathbf{a} = 0.$$

Thus $F(\mathbf{k})$ is a line delta function of strength $2\pi A$ per unit length perpendicular to the vector **a**.

Derived from eqns (2.75), (2.68) and (3.12); or as limiting case of eqn (3.30) $(a \to \infty)$.

Infinite row of delta functions

$$f(\mathbf{r}) = \sum_{n=-\infty}^{+\infty} A\delta(\mathbf{r} - n\mathbf{a}), \qquad F(\mathbf{k}) = \frac{2\pi A}{a} \sum_{m=-\infty}^{+\infty} \delta(\mathbf{k} - \mathbf{K})$$

$$[\mathbf{K} \cdot \mathbf{a} = 2\pi m. \quad m = 0, \pm 1, \pm 2, \ldots]. \qquad (3.35)$$

Thus $F(\mathbf{k})$ represents a set of line delta functions perpendicular to **a**, with spacing $2\pi/|\mathbf{a}|$ and strength $(2\pi A/a)$ per unit length.

Derived from eqns (2.68), (2.72) and (3.12).

Infinite lattice of delta functions

$$f(\mathbf{r}) = A \sum_{n_1, n_2} \delta(\mathbf{r} - n_1\mathbf{a} - n_2\mathbf{b}) \qquad [n_1, n_2 = 0, \pm 1, \pm 2 \ldots]$$

$$F\mathbf{k} = (2\pi)B \sum_{m_1, m_2} \delta(\mathbf{k} - m_1\mathbf{a}^* - m_2\mathbf{b}^*) \quad [m_1, m_2 = 0, \pm 1, \pm 2, \ldots].$$

$$(3.36)$$

$$\frac{A}{B} = \frac{|\mathbf{a} \times \mathbf{b}|}{2\pi} = \frac{2\pi}{|\mathbf{a}^* \times \mathbf{b}^*|}$$

$$\mathbf{a}^* \cdot \mathbf{a} = 2\pi \qquad \mathbf{b}^* \cdot \mathbf{b} = 2\pi$$

$$\mathbf{a}^* \cdot \mathbf{b} = \mathbf{b}^* \cdot \mathbf{a} = 0$$

$$\mathbf{a}^* = 2\pi \frac{\mathbf{b} \times (\mathbf{a} \times \mathbf{b})}{(\mathbf{a} \times \mathbf{b})^2} \qquad \mathbf{b}^* = 2\pi \frac{\mathbf{a} \times (\mathbf{b} \times \mathbf{a})}{(\mathbf{a} \times \mathbf{b})^2}$$

$$\mathbf{a} = 2\pi \frac{\mathbf{b}^* \times (\mathbf{a}^* \times \mathbf{b}^*)}{(\mathbf{a}^* \times \mathbf{b}^*)^2} \qquad \mathbf{b} = 2\pi \frac{\mathbf{a}^* \times (\mathbf{b}^* \times \mathbf{a}^*)}{(\mathbf{a}^* \times \mathbf{b}^*)^2}$$

$f(\mathbf{r})$ and $F(\mathbf{k})$ both represent lattices of delta functions, each of which is the reciprocal lattice of the other, as shown in Fig. 3.1.

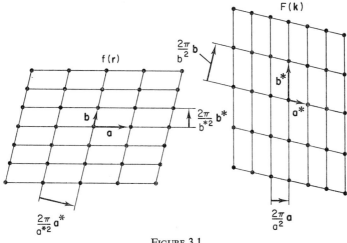

FIGURE 3.1

Derived from eqn (3.35) by convoluting two rows of delta functions to form a lattice. The lattice in $\mathbf{k}$ space results from the product of two sets of line delta functions.

Modified lattice

$$f(\mathbf{r}) = Ah(\mathbf{r}) \Sigma g(\mathbf{r} - n_1\mathbf{a} - n_2\mathbf{b})$$

$$F(\mathbf{k}) = (B/2\pi) \Sigma G(m_1\mathbf{a}^* + m_2\mathbf{b}^*)H(\mathbf{k} - m_1\mathbf{a}^* - m_2\mathbf{b}^*) \quad (3.37)$$

$[A, B, \mathbf{a}, \mathbf{b}, \mathbf{a}^*$ and $\mathbf{b}^*$ related as in eqn (3.36) above].

This result may be stated in words: if the point delta functions in eqn (3.36) above are each replaced by some localised function $g(\mathbf{r})$, and if the whole lattice is modulated or truncated by some function $h(\mathbf{r})$, then in the transform the point delta functions must each be replaced by $H(\mathbf{k})$ and the whole modulated by $G(\mathbf{k})$.

The result follows from the combined application of the product and convolution theorems to the result (3.36) above.

3.8   Table of Three-Dimensional Transforms

$$F(\mathbf{k}) = \iiint e^{-i\mathbf{k}\cdot\mathbf{r}} f(\mathbf{r}) \, d\mathbf{r} = FT^- f(\mathbf{r})$$

$$f(\mathbf{r}) = \left(\frac{1}{2\pi}\right)^3 \iiint e^{+i\mathbf{k}\cdot\mathbf{r}} F(\mathbf{k}) \, d\mathbf{k} = FT^+ F(\mathbf{k})$$

$$FT^-\{F(\mathbf{k})\} = (2\pi)^3 f(-\mathbf{r}). \qquad FT^+\{f(\mathbf{r})\} = \frac{F(-\mathbf{k})}{(2\pi)^3}.$$

Gaussian function

$$f(\mathbf{r}) = A\, e^{-a^2 r^2}$$

$$F(\mathbf{k}) = \frac{A\pi^{3/2}}{a^3}\, e^{-k^2/4a^2}. \tag{3.38}$$

Derived from eqns (2.38) and (3.12).

Gaussian with complex characteristic

$$f(\mathbf{r}) = A\, e^{\pm ia^2 r^2}$$

$$F(\mathbf{k}) = -\frac{A(1 \mp i)(2\pi)^{3/2}}{4a^3}\, e^{\mp ik^2/4a^2}. \tag{3.39}$$

Derived from eqn (3.38) with $a^2 \to \pm ia^2$.

Exponential

$$f(\mathbf{r}) = \frac{1}{r}\, e^{-ar}$$

$$F(\mathbf{k}) = \frac{4\pi}{k^2 + a^2}. \tag{3.40}$$

Derived using eqn (3.18) followed by a standard integration.

Sphere of radius R

$$f(\mathbf{r}) = A \qquad [|r| \leqslant R]$$

$$= 0 \qquad \text{otherwise.} \tag{3.41}$$

$$F(\mathbf{k}) = \frac{4\pi R^3 A}{3} \left\{ \frac{3(\sin kR - kR\cos kR)}{(kR)^3} \right\}.$$

Derived using eqn (3.18) followed by a standard integration.

Spherical shell of radius R of total strength A

$$f(\mathbf{r}) = \frac{A}{4\pi R^2 a} \qquad [R \leqslant r \leqslant (R + a);\ a \ll R]$$

$$= 0 \qquad \text{otherwise.} \tag{3.42}$$

$$F(\mathbf{k}) = \frac{A \sin kR}{kR}.$$

Derived using eqn (3.18) followed by delta function integration.

Delta function at $\mathbf{r} = \mathbf{r}_0$

$$f(\mathbf{r}) = A\delta(\mathbf{r} - \mathbf{r}_0)$$

$$F(\mathbf{k}) = A\,e^{-i\mathbf{k}\cdot\mathbf{r}_0}$$

(3.43)

Derived from eqns (2.69) and (3.12), or verify by delta function integration.

Line delta function in direction $\mathbf{a}$; *strength A per unit length*

$$f(\mathbf{r}) = A\delta(\mathbf{r} - \alpha\mathbf{a}) \qquad [-\infty < \alpha < +\infty]$$

$$F(\mathbf{k}) = 2\pi A\delta(\mathbf{k} - \mathbf{K}) \qquad [\mathbf{K}\cdot\mathbf{a} = 0].$$

(3.44)

$F(\mathbf{k})$ represents a plane through the origin perpendicular to $\mathbf{a}$, of strength $2\pi A$ for unit area. Derived from eqn (3.12) using eqns (2.68) and (2.75).

Infinite row of delta functions

$$f(\mathbf{r}) = A\sum_n \delta(\mathbf{r} - n\mathbf{a}) \qquad [n = 0, \pm1, \pm2, \ldots]$$

$$F(\mathbf{k}) = \frac{2\pi A}{a}\delta(\mathbf{k} - \mathbf{K}) \qquad [\mathbf{K}\cdot\mathbf{a} = 2\pi m;\; m = 0, \pm1, \pm2, \ldots]$$

(3.45)

$F(\mathbf{k})$ represents a set of planes perpendicular to $\mathbf{a}$, of spacing $2\pi/a$ and strength $2\pi A/a$ per unit area.
Derived from eqns (2.72) and (2.68) using eqn (3.12).

Plane lattice of points

$$f(\mathbf{r}) = A\sum_{n_1,n_2} \delta(\mathbf{r} - n_1\mathbf{a} - n_2\mathbf{b}) \qquad [n_1, n_2 = 0, \pm1, \pm2, \ldots]$$

$$F(\mathbf{k}) = 2\pi B\sum_{m_1 m_2} \delta(\mathbf{k} - m_1\mathbf{a}^* - m_2\mathbf{b}^* - \alpha(\mathbf{a}^* \times \mathbf{b}^*))$$

(3.46)

$$[-\infty < \alpha < +\infty;\; m_1, m_2 = 0, \pm1, \pm2, \ldots]$$

$$\frac{B}{A} = \frac{2\pi}{|\mathbf{a} \times \mathbf{b}|} = \frac{|\mathbf{a}^* \times \mathbf{b}^*|}{2\pi}$$

$\mathbf{a}$, $\mathbf{b}$, $\mathbf{a}^*$ and $\mathbf{b}^*$ related as in eqn (3.36). $F(\mathbf{k})$ represents a set of line delta functions passing through the points of a plane lattice.
Derived from eqn (3.45) by convoluting two rows of point delta functions.

Three-dimensional lattice of points

$$f(\mathbf{r}) = A\Sigma\delta(\mathbf{r} - n_1\mathbf{a} - n_2\mathbf{b} - n_3\mathbf{c}) \qquad [n_1, n_2, n_3 = 0, \pm 1, \pm 2, \dots]$$

$$F(\mathbf{k}) = (2\pi)^{3/2} B\Sigma\delta(\mathbf{k} - m_1\mathbf{a}^* - m_2\mathbf{b}^* - m_3\mathbf{c}^*) \qquad (3.47)$$

$$[m_1\, m_2\, m_3 = 0, \pm 1, \pm 2, \dots]$$

$$\frac{B}{A} = \frac{(2\pi)^{3/2}}{|\mathbf{a}\cdot\mathbf{b}\times\mathbf{c}|} = \frac{|\mathbf{a}^*\cdot\mathbf{b}^*\times\mathbf{c}^*|}{(2\pi)^{3/2}}$$

$$\mathbf{a}^* = 2\pi\frac{\mathbf{b}\times\mathbf{c}}{\mathbf{a}\cdot\mathbf{b}\times\mathbf{c}} \qquad \mathbf{a} = 2\pi\frac{\mathbf{b}^*\times\mathbf{c}^*}{\mathbf{a}^*\cdot\mathbf{b}^*\times\mathbf{c}^*}$$

$$\mathbf{a}\cdot\mathbf{a}^* = \mathbf{b}\cdot\mathbf{b}^* = \mathbf{c}\cdot\mathbf{c}^* = 2\pi$$

$$\mathbf{a}\cdot\mathbf{b}^* = \mathbf{a}\cdot\mathbf{c}^* = \mathbf{b}\cdot\mathbf{a}^* = \mathbf{b}\cdot\mathbf{c}^* = \mathbf{c}\cdot\mathbf{a}^* = \mathbf{c}\cdot\mathbf{b}^* = 0$$

$|\mathbf{a}\cdot\mathbf{b}\times\mathbf{c}|$ = Volume of unit cell in $\mathbf{r}$ space.

$|\mathbf{a}^*\cdot\mathbf{b}^*\times\mathbf{c}^*|$ = Volume of unit cell in $\mathbf{k}$ space.

Further relations between $\mathbf{a}$, $\mathbf{b}$, $\mathbf{c}$ and $\mathbf{a}^*$, $\mathbf{b}^*$, $\mathbf{c}^*$ are obtained by cyclic interchange.

Derived from eqns (3.45) and (3.46) by convoluting the plane lattice with a row of points.

Modified Lattice

$$f(\mathbf{r}) = Ah(\mathbf{r}) \sum_{n_1 n_2 n_3} g(\mathbf{r} - n_1\mathbf{a} - n_2\mathbf{b} - n_3\mathbf{c})$$

$$F(\mathbf{k}) = (2\pi)^{-3/2} B \sum_{m_1 m_2 m_3} H(\mathbf{k} - m_1\mathbf{a}^* - m_2\mathbf{b}^* - m_3\mathbf{c}^*)$$

$$\times\, G(m_1\mathbf{a}^* + m_2\mathbf{b}^* + m_3\mathbf{c}^*) \qquad (3.48)$$

$$[n_1, n_2, n_3, m_1, m_2, m_3 = 0, \pm 1, \pm 2, \dots].$$

A, B, $\mathbf{a}$, $\mathbf{b}$, $\mathbf{c}$, $\mathbf{a}^*$, $\mathbf{b}^*$, $\mathbf{c}^*$ related as in eqn (3.47) above.

Thus if the point delta functions in eqn (3.47) above are each replaced by some localised function $g(\mathbf{r})$, and if the whole lattice is modulated or truncated by some function $h(\mathbf{r})$, then in the transform the point delta functions must each be replaced by $H(\mathbf{k})$ and the whole truncated or modulated by $G(\mathbf{k})$.

The result follows from the combined application of the product and convolution theorems to the result (3.47) above.

Finite row of points

$$f(\mathbf{r}) = A\Sigma\delta(\mathbf{r} - n\mathbf{a})$$

$$[n = 0, \pm 1, \pm 2, \dots \pm (N-1)/2 \quad \text{for} \quad N \text{ odd}$$

$$= \pm\tfrac{1}{2}, \pm\tfrac{3}{2}, \dots \pm (N-1)/2 \quad \text{for} \quad N \text{ even}]$$

$$F(\mathbf{k}) = \frac{A \sin (N\mathbf{k} \cdot \mathbf{a}/2)}{\sin (\mathbf{k} \cdot \mathbf{a}/2)}. \tag{3.49}$$

$F(\mathbf{k})$ represents an infinite set of 'fuzzy' planes perpendicular to $\mathbf{a}$, of spacing $2\pi/a$.

Derived from eqn (2.71) using eqn (3.12).

Finite plane lattice of points

$$f(\mathbf{r}) = A\Sigma\delta(\mathbf{r} - n_1\mathbf{a} - n_1\mathbf{b})$$

$$[n_1 = 0, \pm 1, \pm 2, \dots, \pm (N_1 - 1)/2 \quad \text{for} \quad N_1 \text{ odd}$$

$$= \pm\tfrac{1}{2}, \pm\tfrac{3}{2}, \dots, \pm (N_1 - 1)/2 \quad \text{for} \quad N_1 \text{ even. Similarly for } n_2]$$

$$F(\mathbf{k}) = A \left\{ \frac{\sin (N_1\mathbf{k} \cdot \mathbf{a}/2)}{\sin (\mathbf{k} \cdot \mathbf{a}/2)} \frac{\sin (N_2\mathbf{k} \cdot \mathbf{b}/2)}{\sin (\mathbf{k} \cdot \mathbf{b}/2)} \right\}. \tag{3.50}$$

$F(\mathbf{k})$ represents an infinite array of parallel, 'fuzzy' lines.

Derived from eqn (3.49) by convoluting two rows of point delta functions to form a lattice.

Finite three-dimensional lattice of points

$$f(\mathbf{r}) = A\Sigma\delta(\mathbf{r} - n_1\mathbf{a} - n_2\mathbf{b} - n_3\mathbf{c})$$

$$[n_1 = 0, \pm 1, \pm 2, \dots, \pm (N_1 - 1)/2 \quad \text{for} \quad N_1 \text{ odd}$$

$$= \pm\tfrac{1}{2}, \pm\tfrac{3}{2} \dots \pm (N_1 - 1)/2 \quad \text{for} \quad N_1 \text{ even.}$$

$$\text{Similarly for } n_2, n_3 \text{ and } N_2, N_3].$$

$$F(\mathbf{k}) = A \left\{ \frac{\sin (N_1\mathbf{k} \cdot \mathbf{a}/2)}{\sin (\mathbf{k} \cdot \mathbf{a}/2)} \frac{\sin (N_2\mathbf{k} \cdot \mathbf{b}/2)}{\sin (\mathbf{k} \cdot \mathbf{b}/2)} \frac{\sin (N_3\mathbf{k} \cdot \mathbf{c}/2)}{\sin (\mathbf{k} \cdot \mathbf{c}/2)} \right\}. \tag{3.51}$$

$F(\mathbf{k})$ represents an infinite lattice of 'fuzzy' points.

Derived by convoluting a finite lattice of points (3.50) with a finite row of points (3.49) using eqn (3.22).

Chapter 4

Energy Spectra and Power Spectra

4.1 Introduction

If the pressure deviation in a plane sound wave varies with time as $f(t)$ then the instantaneous rate of energy transfer across unit area may be shown, to a good approximation, to be proportional to $f^2(t)$. For a periodic wave-form the mean energy flow will thus be proportional to the mean value $\langle f^2(t) \rangle$ with the average taken over one period. If we write $f(t)$ as the Fourier series

$$\sum_{n=0}^{\infty} \{a_n \cos n\omega t + b_n \sin n\omega t\}$$

we are regarding the wave as equivalent to a superposition of waves of various frequencies and it is natural to consider each frequency component as carrying a portion of the energy. We tentatively expect the energy flow due to the nth harmonic to be proportional to $\langle \{a_n \cos n\omega t + b_n \sin n\omega t\}^2 \rangle$ i.e. proportional to $\frac{1}{2}(a_n^2 + b_n^2)$. This apportioning of the energy between the harmonics is made especially plausible by the following relation, which is readily verified by using the expansion for $f(t)$ and noting that appropriate cross-terms average to zero.

$$\langle f^2(t) \rangle = a_0^2 + \sum_{n=1}^{\infty} \left\{ \frac{a_n^2 + b_n^2}{2} \right\}. \tag{4.1}$$

This means that in the case discussed, with $\langle f(t) \rangle = 0$ so that $a_0 = 0$, the sum of the powers apportioned tentatively as above, does indeed add up correctly to the total power of the wave.

Whether or not the power is 'really' divided amongst the harmonics in this way is not a question which depends solely on the above result, but depends on a consideration of the context in which the idea is to be tested. For instance if the sound impinges on a damped resonant system, energy is absorbed by this system and one does indeed find that the combined system (sound wave plus absorber) behaves just as though the incident energy is distributed among the various harmonics and that the rate of absorption depends on the frequency range to which the resonant system is 'tuned'. Such an analysis is discussed in detail in Chapters 7 and 8.

Let us now consider $f(t)$ to be aperiodic, the total energy transfer being finite. We now wonder naturally whether the energy transfer may be considered as spread over a continuous range of frequencies. If $f(t)$ is written as the Fourier integral

$$\int_0^\infty \{A(\omega)\cos \omega t + B(\omega)\sin \omega t\}\, d\omega$$

then by analogy with the previous result we tentatively apportion the energy proportionately to $\frac{1}{2}\{A^2(w) + B^2(w)\}$ per unit frequency range. This is indeed plausible on account of Parseval's theorem, eqn (2.27), which can be readily adapted using eqn (2.10) to give:

$$\int_{-\infty}^{+\infty} f(t)f^*(t)\, dt = 2\pi \int_0^\infty \left\{\frac{A(\omega)A^*(\omega) + B(\omega)B^*(\omega)}{2}\right\} d\omega. \qquad (4.2)$$

This means that the total energy transfer over all time does indeed equal the integrated sum over all frequencies of the energy apportioned as tentatively suggested provided a factor 2π is inserted. We have quoted Parseval's theorem in its general complex form to show how our results could be generalised to cover a context in which the energy flow was proportional to the squared modulus of some complex function.

An intuitive approach, such as we have given so far, can lead one astray if relied on too much. Consider for instance the following argument. A bell is struck at time $t = 0$, and near it is placed a frequency sensitive detector tuned to a narrow range of frequencies somewhere within the frequency range covered by the Fourier transform of the pressure fluctuation produced by the bell. Since each component in the Fourier transform is a harmonic wave extending over all time, before $t = 0$ as well as after, we conclude that the detector ought to indicate a response even before the bell is rung. Our dilemma is made clear if we now consider what will happen if we decide not to ring the bell after all! In fact of course the detector will only indicate anything from $t = 0$ onwards, and in a typical detector based on a resonant device the response will rise to a maximum and then fall off again. We leave a detailed discussion until Chapters 7 and 8, but note simply in passing that eqn (4.2) applies to the total energy transfer over all time and in no way implies that the term involving $\{A(\omega)A^*(\omega) + B(\omega)B^*(\omega)\}$ represents a uniform flow of energy.

4.2 Energy Spectrum

With the above preamble we now define the energy spectrum $S_f(y)$ of a function $f(x)$ by:

$$S_f(y) = F(y)F^*(y) \tag{4.3}$$

where

$$F(y) = \int_{-\infty}^{+\infty} e^{-ixy} f(x)\, dx.$$

This definition is consistent with the following relation

$$\int_{-\infty}^{+\infty} S_f(y)\, dy = \int_{-\infty}^{+\infty} F(y)F^*(y)\, dy = 2\pi \int_{-\infty}^{+\infty} f(x)f^*(x)\, dx. \tag{4.4}$$

This relation, which follows directly from Parseval's theorem, eqn (2.27), lies at the root of the concept of energy spectrum since it confirms that the 'total energy' in the spectrum is proportional to the 'total energy' in the parent function. In the literature various other conventions are used in which the energy spectrum is defined alternatively as $2|F(\omega)|^2$, as $(1/2\pi)|F(\omega)|^2$, or as $(1/\pi)|F(\omega)|^2$. We are adopting what seems to be the most widespread convention in applied science.

Despite our use of the word energy our definition of energy spectrum is to be regarded as a purely mathematical definition which applies irrespective of any physical interpretation for $f(x)$ or $F(y)$. Our use of the transform $F(y)$ means that $S_f(y)$ is defined for positive and negative values of y; in a context where y only has physical meaning for positive values, as in the frequency analysis of a sound wave discussed in Section 4.1, then we find that the quantity of direct physical interest is closely related to $S_f(y)$. For instance in the analysis of the sound wave the energy transfer per unit frequency range is

$$2\pi \left\{ \frac{A(\omega)A^*(\omega) + B(\omega)B^*(\omega)}{2} \right\} = \frac{1}{2\pi} \{S_f(\omega) + S_f(-\omega)\} \quad [\omega > 0]. \tag{4.5}$$

This equality is easily checked using the results given in eqns (2.7) and (2.8).

We have tacitly assumed in our definition that $f(x)$ does possess an energy spectrum; a sufficient condition for this is that $f(x)$ should be square integrable so that

$$\int_{-\infty}^{+\infty} f(x)f^*(x)\, dx < \infty. \tag{4.6}$$

Such functions are often called finite energy functions since the total energy transfer in a system is so often proportional to an integral of this sort. Thus although $f(x)$ may be Fourier transformable (see Section 2.3) it does not necessarily possess an energy spectrum. In particular if $F(y)$ includes one or more delta functions then on squaring the modulus to get the energy spectrum according to eqn (4.3) we arrive at the square of a delta function this being a singularity with infinite area which is not readily handled.

4.3 Power Spectrum

There is a large class of functions which although they are not square integrable, and so do not have an energy spectrum, do satisfy the following limit

$$\langle |f(x)|^2 \rangle = \lim_{X \to \infty} \frac{1}{2X} \int_{-X}^{+X} |f(x)|^2\, dx < \infty. \tag{4.7}$$

Such functions are often called finite power signals since a finite mean rate of energy transfer in a system is often represented by such a function. Periodic signals such as sine or cosine functions fall in this category as also, but less obviously, does a step function. Stationary random functions (see Section 6.2) also fall in this class although interestingly they possess no Fourier transform. For functions, satisfying (4.7) we define a quantity called the power spectrum, $P_f(y)$, which plays a role for finite power signals which is analogous to that played by $S_f(y)$ for finite energy signals. In defining $P_f(y)$ it is useful to construct a finite energy signal $f_X(x)$ from $f(x)$ by truncation so that

$$\left.\begin{array}{ll} f_X(x) = f(x) & [|x| \leqslant X] \\[2mm] f_X(x) = 0 & [|x| > X] \end{array}\right\}. \tag{4.8}$$

We now define $P_f(y)$ by:

$$P_f(y) = \lim_{X \to \infty} \frac{1}{2X} F_X(y)F_X^*(y) \tag{4.9}$$

where
$$F_X(y) = \int_{-\infty}^{+\infty} e^{-ixy} f_X(x)\, dx.$$

Just as the integrated value of an energy spectrum over all frequencies is proportional to the total energy transfer in a signal (eqn (4.4)) so we expect that the integrated value of a power spectrum shoud match the mean rate of energy transfer in the parent signal. This is readily shown to be so if we apply Parseval's theorem to the truncated function introduced in eqn (4.8) and proceed as follows using eqns (4.7) and (4.9).

$$\int_{-\infty}^{+\infty} |f_X(x)|^2\, dx = \int_{-\infty}^{+\infty} \frac{1}{2\pi} |F_X(y)|^2\, dy$$

therefore

$$\lim_{X \to \infty} \frac{1}{2X} \int_{-\infty}^{+\infty} |f_X(x)|^2\, dx = \lim_{X \to \infty} \frac{1}{2X} \int_{-\infty}^{+\infty} \frac{1}{2\pi} |F_X(y)|^2\, dy$$

so finally

$$\langle |f(x)|^2 \rangle = \frac{1}{2\pi} \int_{-\infty}^{+\infty} P_f(y)\, dy. \tag{4.10}$$

Care is sometimes necessary in interpreting power spectra, and we will illustrate this using t and ω as variables. $P_f(\omega)$ represents the energy flow per unit frequency range per unit time *averaged over all time*. Only if a suitable local time average is possible can $P_f(\omega)$ be interpreted as a *uniform* flow of energy per unit frequency range per unit time. Such 'smoothing' is possible for a periodic signal and for a stationary random function (see Chapter 6) but it is not possible for instance with a step function. With the step function (equal to zero and unity respectively for negative and positive times) we find that

$$\frac{1}{2\pi} \int_{-\infty}^{+\infty} P_f(\omega)\, d\omega$$

is equal to half the rate of energy flow associated with positive times.

For those cases where the Fourier transform of $f(x)$ consists of delta functions it is useful to be able to derive $P_f(y)$ from $F(y)$ and this can be done using the following relation:

if
$$F(y) = \sum_n a_n \delta(y - y_n)$$

then
$$P_f(y) = \frac{1}{2\pi} \sum_n |a_n|^2 \delta(y - y_n). \tag{4.11}$$

Thus it is correct simply to square the modulae of the delta function amplitudes, as might be guessed, provided we also divide by 2π. This result may be established if we make use of the convolution theorem which will be discussed fully in Chapter 5. First we truncate $f(x)$ to give $f_X(x)$ as in eqn (4.8) by truncating $f(x)$ with a rectangular function equal to unity for $|x| \leqslant X$ and zero otherwise. $F_X(y)$ is then proportional to the convolution of $F(y)$ with the transform of the truncating function. Using eqns (2.34) and (2.48) we have

$$F_X(y) = \frac{1}{2\pi}\left(F(y) \otimes \frac{2\sin Xy}{y}\right) = \sum_n a_n \frac{\sin X(y - y_n)}{\pi(y - y_n)} \qquad (4.12)$$

and

$$P_f(y) = \lim_{X \to \infty} \frac{1}{2X} F_X(y) F_X{}^*(y)$$

$$= \lim_{X \to \infty} \left\{\frac{1}{2X}\sum_n |a_n|^2 \left\{\frac{\sin X(y - y_n)}{\pi(y - y_n)}\right\}^2\right\} \qquad (4.13)$$

$$= \frac{1}{2\pi}\sum_n |a_n|^2 \delta(y - y_n). \qquad (4.14)$$

In deriving eqn (4.14) from eqn (4.13) we have used the expression for a delta function given in eqn (6) of Appendix B. In N dimensions we replace 2π in eqn (4.11) by $2\pi^N$.

4.4 Physical Quantities as the Real Parts of Complex Quantities

It is common to represent a real physical function $f(x)$ as the real part of some complex function, say $m(x)$. It is then important to realize that $S_f(y)$ and $S_m(y)$ are only related simply to each other in special circumstances. In the general case it is simple to show that

if $\qquad\qquad f(x) = \operatorname{Re} m(x) = \tfrac{1}{2}(m(x) + m^*(x))$

$\qquad\qquad\quad f(x) \leftrightarrow F(y)$

$\qquad\qquad\quad m(x) \leftrightarrow M(y) \qquad m^*(x) \leftrightarrow M^*(-y)$

then

$$S_f(y) = \tfrac{1}{4}\{S_m(y) + S_m(-y) + M^*(y)M^*(-y) + M(y)M(-y)\}. \qquad (4.15)$$

For the special case that the last two products in eqn (4.15) are small we have

$$S_f(y) \approx \tfrac{1}{4}\{S_m(y) + S_m(-y)\}. \tag{4.16}$$

This condition is satisfied by a slowly modulated high frequency signal and may be illustrated by considering the case of a damped harmonic oscillation for which

$$\left. \begin{array}{l} f(t) = A \cos \omega_0 t \, e^{-\lambda t} \\ m(t) = A \, e^{i\omega_0 t} \, e^{-\lambda t} \end{array} \right\} \quad [t \geqslant 0]$$

$$f(t) = m(t) = 0 \qquad [t < 0]. \tag{4.17}$$

$S_m(\omega)$ is obtained from eqn (2.45) to be

$$S_m(\omega) = M(\omega)M^*(\omega) = \left\{ \frac{A^2}{\lambda^2 + (\omega - \omega_0)^2} \right\}. \tag{4.18}$$

This is the well known Lorentzian function centred on $\omega = \omega_0$. $S_f(\omega)$ may be obtained in like manner from eqn (2.46) but it is evident that there are awkward cross terms in the product and the answer is not a simple Lorentzian. If however the damping is small so that $\lambda \ll \omega_0$ then $M(\omega)$ is negligible except within a region of width λ centred on $\omega = \omega_0$, and we may use the eqn (4.16) to give

$$S_f(\omega) \approx \frac{A^2}{4} \left\{ \frac{1}{\lambda^2 + (\omega - \omega_0)^2} + \frac{1}{\lambda^2 + (\omega + \omega_0)^2} \right\}. \tag{4.19}$$

This result is of course also derivable direct from eqn (2.46) in this approximation.

We may in fact generalise this example and consider any type of slow modulation of a carrier wave. Suppose a real function $f(t)$ varies as

$$f(t) = \text{Re}\, m(t)$$

where

$$m(t) = n(t)\, e^{i\omega_0 t}.$$

Provided the modulating function $n(t)$ varies slowly compared with an oscillation at frequency ω_0 the approximation of eqn (4.16) holds so that

$$S_f(\omega) \approx \tfrac{1}{4}\{S_m(\omega) + S_m(-\omega)\}.$$

However it follows from the shift theorem, eqn (2.25), that

$$S_m(\omega) = S_n(\omega - \omega_0)$$

and thus

$$S_f(\omega) \approx \tfrac{1}{4}\{S_n(\omega - \omega_0) + S_n(-\omega - \omega_0)\}. \tag{4.20}$$

This allows the spectrum of the signal to be derived from the spectrum of the modulating function. Note that $n(t)$ may be a complex function so that phase modulation as well as amplitude modulation is allowed for in this approximation.

Although the above remarks have been aimed at finite energy signals, finite power signals are treated similarly if we replace energy spectra by power spectra in the above discussion.

4.5   Spectra in Several Dimensions

The formal definitions can clearly be extended in an obvious way to cover systems in many dimensions. As before two types of function can be distinguished, square integrable functions for which

$$\int f(\mathbf{r}) f^*(\mathbf{r}) \, d\mathbf{r} < \infty \tag{4.21}$$

and functions of finite mean squared modulus for which

$$\langle f(\mathbf{r}) f^*(\mathbf{r}) \rangle = \lim_{V \to \infty} \left\{ \frac{1}{V} \int_V f(\mathbf{r}) f^*(\mathbf{r}) \, d\mathbf{r} \right\} < \infty. \tag{4.22}$$

We have used $\mathbf{r}$ to represent an N-dimensional vector, and V an N-dimensional volume. For want of better short descriptive titles we will call these finite energy and finite power functions respectively although the names are clearly not very happy choices when t and ω are no longer the variables.

We define energy and power spectra respectively for the two classes of function by

$$S_f(\mathbf{k}) = F(\mathbf{k}) F^*(\mathbf{k}) \tag{4.23}$$

and

$$P_f(\mathbf{k}) = \lim_{V \to \infty} \frac{1}{V} \left| \int_V e^{-i\mathbf{k} \cdot \mathbf{r}} f(\mathbf{r}) \, d\mathbf{r} \right|^2. \tag{4.24}$$

Parseval's theorem leads to the following generalisations of eqns (4.4) and (4.10):

$$\frac{1}{(2\pi)^N} \int S_f(\mathbf{k}) \, d\mathbf{k} = \int f(\mathbf{r}) f^*(\mathbf{r}) \, d\mathbf{r} \tag{4.25}$$

$$\frac{1}{(2\pi)^N} \int P_f(\mathbf{k}) \, d\mathbf{k} = \langle f(\mathbf{r}) f^*(\mathbf{r}) \rangle. \tag{4.26}$$

Finally if $F(\mathbf{k})$ consists of point delta functions then eqn (4.11) is readily generalised so that if

$$F(\mathbf{k}) = \sum_n a_n \delta(\mathbf{k} - \mathbf{k}_n)$$

then

$$P_f(\mathbf{k}) = \frac{1}{(2\pi)^N} \sum_n |a_n|^2 \delta(\mathbf{k} - \mathbf{k}_n). \qquad (4.27)$$

4.6   Relations Between Functions and Their Spectra

The relations between a function and its Fourier transform have been described in detail in Sections 2.4 and 3.6 and from these results one can of course derive the relationship between a function and its energy or power spectrum. We will here simply point out what seem to be particularly useful or important results.

Whereas a function is uniquely determined by a knowledge of its Fourier transform and vice versa, it is not true that a function is determined by a knowledge of its spectrum (although the spectrum is determined by a knowledge of the function). A given spectrum $S_f(y)$ may correspond to an infinite set of functions $f(x)$ as is clearly seen by noting that $S_f(y)$ is unaltered if we multiply $F(y)$ by $e^{ig(y)}$ where $g(y)$ is any real function of y. This is equivalent to shifting the phases of the various harmonic components of $f(x)$ and means that an energy or power spectrum is 'insensitive to phase'. This lack of uniqueness is often a difficulty in experimental work when information on a quantity $f(x)$ is available through a technique which gives $S_f(y)$ directly but not $F(y)$.

A special case of the above result is exemplified in the fact that shifting $f(x)$ has no effect on the spectrum as is easily verified using eqn (2.24). This result lies behind the familiar facts that the energy spectrum of a time-dependent signal does not depend on when the signal occurs and that the angular distribution of radiation diffracted by an aperture is not altered by lateral movement of the aperture.

An energy or power spectrum is always real by definition and in addition will possess even symmetry, so that $S_f(y) = S_f(-y)$, if $f(x)$ is either pure real, pure imaginary, of even symmetry, or of odd symmetry.

The energy or power spectrum of a sum of functions is not in general equal to the sum of the individual spectra. Indeed it is easy to verify that

if $f(x) = g(x) + h(x)$

so that $F(y) = G(y) + H(y)$

then

$$S_f(y) = F(y)F^*(y) = S_g(y) + S_h(y) + G(y)H^*(y) + G^*(y)H(y). \qquad (4.28)$$

The cross terms in eqn (4.28) which destroy simple additivity receive interpretation as 'interference' terms in diffractive phenomena, or as indicating a cross-correlation (see Chapter 5) between $g(x)$ and $h(x)$ in other contexts.

The spectrum of a product of two functions, again, is not simply related to the spectra of the individual functions. The convolution theorem (eqn (2.34)) does *not* for instance imply that the final spectrum is the convolution of the individual spectra; instead, for finite energy signals, it states no more than the following

if $$f(x) = g(x)h(x)$$

then $$S_f(y) = \frac{1}{(2\pi)^2} |G(y) \otimes H(y)|^2. \tag{4.29}$$

A simplifying condition arises if $G(y)$ consists of a set of unit delta functions and if $H(y)$ is a function narrow compared with the spacing between the delta functions. It is then straightforward to verify that

$$S_f(y) = \frac{1}{(2\pi)^2} \{S_h(y) \otimes G(y)\}. \tag{4.30}$$

This result applies for instance to a diffraction grating when $g(x)$ describes an infinite grating of fine rulings and $h(x)$ is a truncating function reducing the finite grating to a large but finite one (for a more detailed discussion see Chapter 9).

In contrast to what has just been said the spectrum of the convolution of two functions *does* bear a simple relation to the individual spectra, being equal to their product. Thus for finite energy signals, eqn (2.35) leads directly to the following:

if $$f(x) = g(x) \otimes h(x)$$

then $$S_f(y) = S_g(y)S_h(y). \tag{4.31}$$

In the limit that $g(x)$ and $f(x)$ become signals of infinite energy but finite power, whilst $h(x)$ remains of finite energy, it follows readily from eqn (4.31) using the ideas leading up to eqn (4.9) that

$$P_f(y) = P_g(y)S_h(y). \tag{4.32}$$

These equations (4.31) and (4.32) may be illustrated by returning to consideration of a diffraction grating; if $g(x)$ is related to a grating of fine rulings then $h(x)$ can be used to represent the effect of making the individual rulings broad instead of narrow. The final angular distribution function of diffracted intensity is proportional to the product of the functions appropriate respectively to a grating of fine rulings and a single wide ruling.

Chapter 5

Convolution and Correlation

5.1 Convolution (Faltung or Folding). Definition and Meaning

The convolution (in German 'faltung' which translates as 'folding') of two functions $f(x)$ and $g(x)$ is defined as

$$f(x) \otimes g(x) = \int_{-\infty}^{+\infty} f(x_1)g(x - x_1) \, dx_1. \tag{5.1}$$

In two or more dimensions the definition is extended in an obvious fashion; for instance in three dimensions we have the following, with the integration covering all space:

$$f(\mathbf{r}) \otimes g(\mathbf{r}) = \iiint f(x_1, y_1, z_1)g(x - x_1, y - y_1, z - z_1) \, dx_1 \, dy_1 \, dz_1$$

$$= \iiint f(\mathbf{r}_1)g(\mathbf{r} - \mathbf{r}_1) \, d\mathbf{r}_1. \tag{5.2}$$

Interchanging the two functions has no effect on the convolution so that

$$f(x) \otimes g(x) = g(x) \otimes f(x). \tag{5.3}$$

This is easily verified by changing the variable in eqn (5.1) from x_1 to x_2 using the substitution $x_1 = x - x_2$.

Convoluting two functions is a way of combining them together, and very often the effect is that of 'broadening' the one by the other. We describe below some common situations in which the process of convolution is applicable. These illustrations are not intended as rigorous discussions of the phenomena (many of which will be dealt with in part II of the book) but are intended simply to give a feeling for the meaning of convolution.

Consider first a well collimated beam of light with an angular distribution of intensity which is approximated by a delta function $\delta(\theta)$. If the beam passes through a ground glass screen it is broadened out into a diffuse beam with an angular distribution represented, say, by $B(\theta)$. Suppose now that an incident beam possessing an angular distribution $A(\theta)$ is passed through the screen, then the angular width of the emerging beam will be broader than $A(\theta)$ by an amount depending on $B(\theta)$. A little thought will suffice to show that the angular distribution of the transmitted beam is given by:

$$C(\theta) = \int_{-\pi/2}^{+\pi/2} A(\theta_1)B(\theta - \theta_1)\,d\theta_1$$

$$= A(\theta) \otimes B(\theta). \tag{5.4}$$

The argument may be summarised, in somewhat imprecise terms, by saying that the component of radiation emerging at angle θ due to light incident at angle θ_1 (having been deviated through an angle $\theta - \theta_1$) has an intensity $A(\theta_1)B(\theta - \theta_1)$. In the absence of interference effects the total intensity emerging in a direction θ is thus the integral of this product over all values of θ_1, as in eqn (5.4). Note that the order of the functions is immaterial since we could have argued equally well that the intensity emerging at angle θ due to light incident at $\theta - \theta_1$ (having been deviated through an angle θ_1) has an intensity $A(\theta - \theta_1)B(\theta_1)$. The total intensity is again arrived at by integrating overall incident directions, i.e. over all values of θ_1. We now have

$$C(\theta) = \int_{-\pi/2}^{+\pi/2} B(\theta_1)A(\theta - \theta_1)\,d\theta_1$$

$$= B(\theta) \otimes A(\theta). \tag{5.5}$$

Another example of convolution may be based on the diffraction grating. As a first approximation a diffraction grating is often considered as a set of infinitesimally wide rulings, so that the transmission function may be represented by a set of delta functions. If instead each ruling has a finite width and definite transmission profile then the overall transmission may be regarded as the result of convoluting the set of delta functions with the broadened function characteristic of one slit.

Again the electron density distribution inside a crystal lattice may be regarded as a convolution between a set of delta functions at the lattice points and the local density distribution which is associated with a lattice point.

As a final example consider an instantaneous signal $\delta(t)$ which when applied to some device (such as an electrical network) produces an output signal $g(t)$. We may say that the device has 'blurred' or 'broadened' the signal and converted it from $\delta(t)$ into $g(t)$. By the same arguments as before if we now apply a signal $f(t)$ of finite duration we expect the resulting output to be the convolution of $f(t)$ with $g(t)$.

It is clear from the above examples that the blurring of experimental data by finite instrumental resolution can often be described by means of a convolution of the true data with an instrumental function. The deconvolution of such information is thus of importance, and we discuss a method which is sometimes useful in Section 5.4. The removal of echo from a signal is discussed in Section 10.6, and the removal of blurr from photographs is discussed in Section 12.9.

5.2   Cross-Correlation Function. Definition and Meaning

The cross-correlation function between two functions $f(x)$ and $g(x)$ is a function of x written alternatively as $\rho_{fg}(x)$ or as $f(x) * g(x)$ and defined by

$$\rho_{fg}(x) = f(x) * g(x) = \int_{-\infty}^{+\infty} f^*(x_1)g(x_1 + x)\,dx_1$$

$$= \int_{-\infty}^{+\infty} f^*(x_1 - x)g(x_1)\,dx_1. \qquad (5.6)$$

The equivalence of the two expressions in eqn (5.6) may be readily established by substituting $x_1 + x = x_2$ in the first integral. Other conventions are found in the literature with the cross-correlation defined alternatively as $\int f(x_1)g^*(x_1 + x)\,dx_1$ or as $\int f(x_1)g^*(x_1 - x)\,dx_1$. Unfortunately these definitions are not equivalent.

Equation (5.6) is appropriate if one or other (or both) of the functions is a finite energy signal: if both functions are finite power signals the above expression will in general be infinite and we adopt the modified definition

$$R_{fg}(x) = \lim_{X \to \infty} \frac{1}{2X} \int_{-X}^{+X} f^*(x_1)g(x_1 + x)\,dx_1 \qquad (5.7)$$

$$= \langle f^*(x_1)f(x_1 + x) \rangle. \qquad (5.8)$$

In the expression (5.8) the triangular brackets represent an average over all values of x_1. The definitions just given bear a certain formal similarity to the definition of a convolution (eqn (5.1)), however as we shall see below the use and interpretation are very different. In contrast to the situation with convolution the order in which the functions are taken is important, and the substitution $x_1 = x_2 - x$ in eqns (5.6) and (5.7) readily establishes that

$$\rho_{fg}(x) = \rho^*_{gf}(-x) \tag{5.9}$$

and

$$R_{fg}(x) = R^*_{gf}(-x). \tag{5.10}$$

As the name implies the cross-correlation function is a measure of whether any correlation between two functions exists, and the concept is especially useful and easy to grasp for periodic functions or for certain types of random function associated with noise. Two entirely different and unrelated functions will have zero cross-correlation for all x, whereas two functions associated with physical quantities between which there is a direct causal connection will often have a finite cross-correlation for some or all values of x.

As an illustration let one function $f(t)$ be the deviation from average of the pressure somewhere in a room due to a person speaking, and let the other function $g(t)$ be the corresponding function in another room into which the sound is being relayed (at equal intensity) by a loudspeaker system. Clearly $f(t)$ and $g(t)$ are related, but not perhaps equal due to imperfections in the relay system: both functions are real so we will drop the complex conjugate symbol in what follows. For small delay times (short compared with the period of the highest frequency components present in the sound) the cross-correlation function will have a finite positive value since $g(t_1 + t)$ will be nearly equal to $f(t_1)$ for all values of t_1, and the product $f(t_1)g(t_1 + t)$ will always be positive (or zero). For t equal to about half the average period of oscillation $g(t_1 + t)$ will, for most values of t_1, have the opposite sign from $f(t_1)$ so that the cross-correlation becomes negative. As the delay t increases the cross-correlation function oscillates between positive and negative values. However for large values of t (greater than a minute say) the cross-correlation function will have tended to zero since there is no correlation between $g(t_1 + t)$ and $f(t_1)$ and the product $f(t_1)g(t_1 + t)$ for arbitrary t_1 is just as likely to be positive as negative and the average is zero. Thus we see that as t increases from zero the cross-correlation exhibits oscillations which die away in amplitude over a time scale equal to the length of a typical vowel or consonant sound. Strictly on account of the time delay in transmission from one room to the other the peak in the correlation function will be offset from zero by an amount equal to the delay. If the relay system introduces random noise which is superimposed on the relayed sound the cross-correlation function is unaltered. This may be seen by replacing the function $g(t)$ in the above discussion by $g(t) + h(t)$ where

$h(t)$ represents the noise. In considering a product such as

$$f(t_1)\{g(t_1 + t) + h(t_1 + t)\}$$

the product involving $f(t_1)$ and $h(t_1 + t)$ is just as likely to be positive as negative for arbitrary t_1 and this term averages to zero leaving our previous results unchanged. If the relay system distorts the sound, then the cross-correlation function will be altered, as will also be the case if the second room possesses an echo or reverberation. These comments will apply equally well if the experiment is of long but finite duration (say 30 minutes) or if it is hypothetically of infinite duration, provided we understand the cross-correlation to refer to $\rho_{fg}(t)$ and $R_{fg}(t)$ respectively for the two cases.

A special case of some interest arises if one function is arranged to be a harmonically varying reference signal of controllable frequency. The cross-correlation function between a finite energy signal $f(t)$ and this reference signal is then a harmonically oscillating function whose amplitude is related to the Fourier transform of $f(t)$. This is shown for a complex oscillatory function as follows:

if
$$g(t) = e^{i\omega_0 t}$$

and
$$f(t) \leftrightarrow F(\omega)$$

then
$$\rho_{gf}(t) = \int e^{-i\omega_0(t_1 - t)} f(t_1) \, dt_1$$

$$= e^{i\omega_0 t} F(\omega_0). \tag{5.11}$$

The definitions in eqns (5.6), (5.7) and (5.8) can clearly be extended to two or more dimensions, so that for instance in three dimensions:

$$\rho_{fg}(\mathbf{r}) = \iiint f^*(x_1 y_1 z_1) g(x_1 + x, y_1 + y, z_1 + z) \, dx_1 \, dy_1 \, dz_1 \tag{5.12}$$

$$= \iiint f^*(\mathbf{r}_1) g(\mathbf{r}_1 + \mathbf{r}) \, d\mathbf{r}_1 \tag{5.13}$$

$$R_{fg}(\mathbf{r}) = \lim_{X,Y,Z \to \infty} \frac{1}{2X2Y2Z} \int_{-X}^{X} \int_{-Y}^{Y} \int_{-Z}^{Z} f^*(x_1 y_1 z_1) \tag{5.14}$$

$$\times g(x_1 + x, y_1 + y, z_1 + z) \, dx_1 \, dy_1 \, dz_1 \tag{5.15}$$

$$= \lim_{V \to \infty} \frac{1}{V} \iiint_V f^*(\mathbf{r}_1) g(\mathbf{r}_1 + \mathbf{r}) \, d\mathbf{r}_1 \tag{5.16}$$

$$= \langle f^*(\mathbf{r}_1) g(\mathbf{r}_1 + \mathbf{r}) \rangle.$$

5.3   Autocorrelation Function. Definition and Meaning

The autocorrelation function of a function $f(x)$ is simply a special case of the cross-correlation defined in eqns (5.6), (5.7) and (5.8), with $f(x) = g(x)$. Thus:

$$\rho_f(x) = \int_{-\infty}^{+\infty} f^*(x_1)f(x_1 + x)\,dx_1 \tag{5.17}$$

$$R_f(x) = \lim_{X \to \infty} \frac{1}{2X}\int_{-X}^{+X} f^*(x_1)f(x_1 + x)\,dx_1 \tag{5.18}$$

$$= \langle f^*(x_1)f(x_1 + x)\rangle. \tag{5.19}$$

Note that eqn (5.17) will be applicable if $f(x)$ is a finite energy function, and eqns (5.18) and (5.19) if it is a finite power function.

The autocorrelation function is a measure of the correlation between the values of a function at values of the variable differing by an amount x. It is especially useful in characterising the properties of irregular functions. As an illustration of the use of an autocorrelation function let us consider an irregular porous or powered material whose local density varies with position according to the function $f(\mathbf{r})$. Consider a straight line running through the sample which we shall treat as an x axis, with origin somewhere well within the sample, and let us for the moment consider simply the density $f(x)$ at points along this line. Clearly from the de inition, eqn (5.17),

$$\rho_f(0) = \int_{-\infty}^{+\infty} f^*(x_1)f(x_1)\,dx_1 \tag{5.20}$$

$$= L\langle f f^*\rangle_L \tag{5.21}$$

so that for intervals $x = 0$ the autocorrelation function is simply related to the mean square density within the sample (L is the length of the sample intercepting our axis, and $\langle\ \rangle_L$ represents an average taken over this length). For increasing intervals x, $\rho_f(x)$ will decrease and tend towards $L|\langle f\rangle|^2$ at values of x large compared with pore or particle size yet still small compared with L. This can be seen by writing $f(x)$ as the sum of its mean value f_0 and a fluctuating deviation from this mean $f_1(x)$ so that

$$\rho_f(x) = \int_{-\infty}^{+\infty} \{f_0 + f_1(x_1)\}^*\{f_0 + f_1(x_1 + x)\}\,dx_1 \tag{5.22}$$

$$= \int_{-\infty}^{+\infty} f_0^* f_0\,dx_1 = Lf_0^* f_0 \quad \text{[Large } x\text{].} \tag{5.23}$$

In deriving (5.23) from (5.22) the cross terms other than $f_0^* f_0$ average to zero since for arbitrary x_1, $f_1(x_1)$ and $f_1(x_1 + x)$ are independent and just as likely to be positive as negative. For yet larger values of x, larger than the dimension L of the sample, $\rho_f(x)$ clearly tends to zero since one or other of the terms in the product $f^*(x_1)f(x_1 + x)$ must be zero. The shape of the autocorrelation function thus summarises statistical information on the various characteristic lengths that describe the specimen. If there is any local regularity in the parent function this will show up as an oscillation in the value of the autocorrelation function. Thus if the powdered specimen consisted of roughly equally sized particles packed together then over distances of about 5 to 10 diameters the density will oscillate with a periodicity roughly equal to that of the particle diameter. Due to the randomness of the packing the strict periodicity will be lost for distances much greater than this. The autocorrelation function can be drawn schematically as below to illustrate this discussion, the region for large x being dashed to indicate a compressed scale.

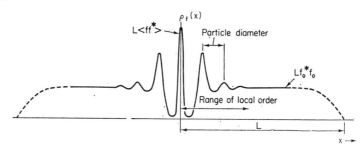

FIGURE 5.1

The extension to two or more dimensions can be made in a natural way so that for instance in three dimensions:

$$\rho_f(\mathbf{r}) = \iiint f^*(x_1 y_1 z_1) f(x_1 + x, y_1 + y, z_1 + z)\, dx_1\, dy_1\, dz_1 \quad (5.24)$$

$$= \iiint f^*(\mathbf{r}_1) f(\mathbf{r}_1 + \mathbf{r})\, d\mathbf{r}_1 \quad (5.25)$$

$$R_f(\mathbf{r}) = \lim_{X,Y,Z \to \infty} \frac{1}{2X,\, 2Y,\, 2Z} \int_{-X}^{+X}\int_{-Y}^{+Y}\int_{-Z}^{+Z} f_1^*(x_1 y_1 z_1)$$
$$\times f(x_1 + x, y_1 + y, z_1 + z)\, dx_1\, dy_1\, dz_1 \quad (5.26)$$

$$= \lim_{V \to \infty} \frac{1}{V} \iiint_V f^*(\mathbf{r}_1) f(\mathbf{r}_1 + \mathbf{r})\, d\mathbf{r}_1 \quad (5.27)$$

$$= \langle f^*(\mathbf{r}_1) f(\mathbf{r}_1 + \mathbf{r}) \rangle \quad (5.28)$$

5.4   The Convolution and Wiener–Khintchine Theorems

These two closely related theorems are of fundamental importance in that they link together the concepts of convolution, correlation and energy spectrum by means of Fourier transform techniques.

The convolution theorem (proved in Appendix G) states that "the Fourier transform of a convolution of two functions is proportional to the product of the individual transforms and conversely the Fourier transform of a product of two functions is proportional to the convolution of their individual transforms." More precisely it states that

if $f(x) \leftrightarrow F(y)$

and $g(x) \leftrightarrow G(y)$

then $$f(x)g(x) \leftrightarrow \frac{1}{2\pi} F(y) \otimes G(y) \qquad (5.29)$$

and $$f(x) \otimes g(x) \leftrightarrow F(y)G(y). \qquad (5.30)$$

Clearly one very important use of the convolution theorem lies in obtaining the Fourier transform of a complicated function from the Fourier transforms of simpler functions when the complicated function can be compounded from simpler ones by convolution or multiplication. We have already made great use of this in Section (2.5). Conversely the theorem sometimes provides a way of obtaining the convolution of two functions by multiplying the Fourier transforms together and then inversely transforming this product. In this way it is easy to verify the two well-known results that two rectangular functions convolute to give a triangular function (using eqns (2.48) and (2.55)), that two Lorentzians convolute to give another Lorentzian whose width at half height is equal to the sums of the individual widths (using eqn (2.39)), and that two Gaussians convolute to give another Gaussian whose width squared at $1/e$ height is equal to the sum of the individual squared widths (using eqn (2.38)). Another important use lies in 'deconvoluting' or 'unfolding' two functions as follows. Suppose we know the function $\{f(x) \otimes g(x)\}$ and also the function $g(x)$ and we wish to find $f(x)$. We proceed by Fourier transforming both the known functions giving $F(y)G(y)$ and $G(y)$ respectively; division now gives us $F(y)$ which can finally be inversely transformed to give $f(x)$. This procedure is clearly attractive as a method of allowing for the effects of instrumental blurring of data.

The Wiener–Khintchine theorem (proved in Appendix J) states that the Fourier transform of the autocorrelation function is equal to the energy or power spectrum of a function. More precisely it means that for a finite energy function $f(x)$:

$$\rho_f(x) \leftrightarrow S_f(y) \tag{5.31}$$

whilst for a finite power function we have correspondingly:

$$R_f(x) \leftrightarrow P_f(y). \tag{5.32}$$

Note that eqn (5.32) holds irrespective of whether $f(x)$ possesses a Fourier transform so that the result covers stationary random functions as well as periodic ones. In many treatments of the subject eqns (5.31) and (5.32) are adopted as the definitions of energy and power spectra, and in this case the working in Appendix J serves to prove our eqns (4.3) and (4.9). As is also shown in Appendix J, the Wiener–Khintchine theorem as quoted above is in fact a special case of a more general theorem which states that the Fourier transform of the cross-correlation between two functions is equal to the so-called cross energy spectrum or cross power spectrum.

The Wiener–Khintchine theorem finds frequent application when one is dealing with random functions. In this context it is often either the power spectrum or the auto-correlation function that is experimentally measureable or significant and the theorem allows the one to be related to the other through Fourier transformation. We have already remarked in Section 4.6 that knowledge of an energy or power spectrum does not imply a full knowledge of the 'parent' function. This fact is exemplified here in that Fourier transforming the energy or power spectrum gives us the auto-correlation function which also provides an incomplete specification of the parent function.

The statements of the two theorems given above can be readily extended to two or more dimensions by replacing x and y in eqns (5.31) and (5.32) by appropriate vectors.

5.5 Summary of Relationships

We now collect together the definitions and theorems obtained so far. For the purposes of this summary we introduce a constant A to allow for conversion from one convention to another in respect of factors of 2π and $\sqrt{2\pi}$ which appear in the definitions of Fourier transforms. The convention used in the rest of this book is obtained by putting $A = 1$. Symmetry between the transform and inverse transform is obtained with $A = (2\pi)^{-\frac{1}{2}}$, and a further convention sometimes used corresponds to $A = (2\pi)^{-1}$. The results may be converted to n-dimensional relationships simply by replacing 2π by $(2\pi)^n$ and replacing x and y by n-dimensional vectors.

Definitions

$$F(y) = FT^- f(x) = A \int e^{-ixy} f(x) \, dx$$

$$FT^- F(y) = A \int e^{-ixy} F(y) \, dy$$

$$FT^+ f(x) = \frac{1}{2\pi A} \int e^{+ixy} f(x) \, dx \qquad (5.33)$$

$$f(x) = FT^+ F(y) = \frac{1}{2\pi A} \int e^{+ixy} F(y) \, dy$$

$$f(x) \otimes g(x) = \int f(x_1) g(x - x_1) \, dx_1 \qquad (5.34)$$

$$\rho_{fg}(x) = f(x) * g(x) = \int f^*(x_1) g(x_1 + x) \, dx_1 \qquad (5.35)$$

$$\rho_f(x) = f(x) * f(x) = \int f^*(x_1) f(x_1 + x) \, dx_1. \qquad (5.36)$$

Fourier's Theorem

If
$$F(y) = FT^- f(x)$$

then
$$f(x) = FT^+ F(y) \qquad (5.37)$$

that is
$$f(x) \leftrightarrow F(y)$$

$$FT^+ FT^- f(x) = FT^- FT^+ f(x) = f(x) \qquad (5.38)$$

$$FT^- F(y) = FT^- FT^- f(x) = 2\pi A^2 f(-x) \qquad (5.39)$$

$$FT^+ f(x) = FT^+ FT^+ F(y) = \frac{1}{2\pi A^2} F(-y) \qquad (5.40)$$

$$f^*(x) \leftrightarrow F^*(-y) \qquad (5.41)$$

$$f^*(-x) \leftrightarrow F^*(y) \qquad (5.42)$$

$$f(-x) \leftrightarrow F(-y). \qquad (5.43)$$

Parseval's theorem

$$\int f(x) g^*(x) \, dx = \frac{1}{2\pi A^2} \int F(y) G^*(y) \, dy. \qquad (5.44)$$

Convolution and Wiener–Khintchine theorems

$$f(x) \otimes g(x) \leftrightarrow \frac{1}{A} F(y)G(y) \tag{5.45}$$

$$f(x)g(x) \leftrightarrow \frac{1}{2\pi A} F(y) \otimes G(y) \tag{5.46}$$

$$f(x) * g(x) \leftrightarrow \frac{1}{A} F^*(y)G(y) \tag{5.47}$$

$$f^*(x)g(x) \leftrightarrow \frac{1}{2\pi A} F(y) * G(y). \tag{5.48}$$

Chapter 6

Random Functions

6.1 Introduction

The trace seen on an oscilloscope when electronic noise is displayed typifies one type of random signal; there are however many types of random signal, an irregularly spaced set of delta functions being one other example. Random functions can of course be functions of variables other than time, and can exist in more than one dimension. For simplicity we will restrict our discussion to one dimension; we will also phrase the discussion with time as the variable, in deference to the wide application of our results in communication theory. The adaptations to other contexts are usually obvious.

Once a random signal of finite duration has been recorded as a function of time it is no longer an 'unknown' signal and it is perfectly possible to obtain its Fourier transform and spectrum by numerical methods using the standard definitions. One may well wonder why special attention need be paid to random signals as distinct from any other kind of signal. The answer lies in the fact that very often we do not have a complete record of the function but only know certain average properties associated with it. Our aim then is to obtain relations between such average properties (including the spectrum and autocorrelation function) which will be useful in such situations. The results will of course be equally applicable to a signal that is not random.

One case in which our knowledge is statistical rather than detailed arises when the signal is of very long duration and we have detailed knowledge of only a portion of it. Another case arises when there is an 'ensemble' of functions which are similar in their statistical properties but different in detail and we only possess information derived from a limited number of members of the ensemble.

6.2   Stationary Random Functions

For simplicity we will restrict further discussion to a class of random functions of infinite duration known as stationary random functions. A stationary random function is one whose statistical properties do not vary with time. More precisely such a function is defined as one for which the following limits exist, independent of the choice of T_1:

$$\langle f(t) \rangle = \lim_{T \to \infty} \frac{1}{T} \int_{T_1}^{T_1+T} f(t)\,\mathrm{d}t \qquad (6.1)$$

$$= \lim_{T \to \infty} \frac{1}{T} \int_{-\infty}^{+\infty} f_T(t)\,\mathrm{d}t \qquad (6.2)$$

$$R_f(t) = \langle f^*(t_1) f(t_1 + t) \rangle = \lim_{T \to \infty} \frac{1}{T} \int_{T_1}^{T_1+T} f^*(t_1) f(t_1 + t)\,\mathrm{d}t_1 \qquad (6.3)$$

$$= \lim_{T \to \infty} \frac{1}{T} \int_{-\infty}^{+\infty} f_T^*(t_1) f_T(t_1 + t)\,\mathrm{d}t_1. \qquad (6.4)$$

In these expressions $f_T(t)$ represents a 'chopped off' portion of $_f(t)$ defined by

$$\left. \begin{array}{ll} f_T(t) = f(t) & \text{for} \quad T_1 < t < T_1 + T \\[2mm] f_T(t) = 0 & \text{otherwise.} \end{array} \right\} \qquad (6.5)$$

The expressions (6.1) and (6.2) above define the average value of $f(t)$, the triangular bracket representing an average over all values of t. The expressions (6.3) and (6.4) define the autocorrelation function $R(t)$, the triangular brackets representing averages over all values of t_1. The expressions (6.1) and (6.2) are obviously equivalent, and it requires only a little thought to see that expressions (6.3) and (6.4) are also. The noise in an electronic circuit is an example of a stationary random function (so long as the circuit is not altered), whilst a quantity such as the world population is not.

A stationary random function is clearly a finite power signal rather than a finite energy signal, and accordingly the power spectrum may be defined by

$$P_f(\omega) = \lim_{T \to \infty} \frac{1}{T} F_T^*(\omega) F_T(\omega) \tag{6.6}$$

where

$$f_T(t) \leftrightarrow F_T(\omega)$$

and Parseval's theorem and the Wiener–Khintchine theorem (see eqns (4.10) and (5.32)) give us

$$\int_{-\infty}^{+\infty} P_f(\omega) \, d\omega = 2\pi \left\langle |f(t)|^2 \right\rangle \tag{6.7}$$

and

$$R_f(t) \leftrightarrow P_f(\omega). \tag{6.8}$$

These relations, (6.7) and (6.8) are of crucial importance in dealing with stationary random functions since the autocorrelation function, power spectrum and mean square value are the most used quantities for describing such functions. In contrast it may be noted that we have not as yet referred to the Fourier transform of a stationary random function; this is because a transform defined in the usual way does not exist, or more precisely because

$$\int_{-T}^{+T} e^{-i\omega t} f(t) \, dt \to \infty \quad \text{as} \quad T \to \infty. \tag{6.9}$$

An attempt at normalising this per unit time is also unsuccessful since we find that

$$\frac{1}{2T} \int_{-T}^{+T} e^{-i\omega t} f(t) \, dt \to 0 \quad \text{as} \quad T \to \infty \ [\omega \neq 0]. \tag{6.10}$$

In this context an alternative approach based on ensemble averages is instructive. Consider an ensemble of functions each of finite duration, $f_1(t), f_2(t) \ldots f_i(t) \ldots$, taken say from different oscilloscope traces of electronic noise. Each may be individually transformed so that we obtain an ensemble of functions, $F_1(\omega), F_2(\omega) \ldots F_i(\omega) \ldots$. The important point is that if we average these functions $F_i(\omega)$ then as the ensemble becomes larger and larger so the average tends towards the transform of the corresponding average of the functions $f_i(t)$. This result follows directly from the additivity of transforms (eqn (2.20)). For a stationary random function of zero mean value the average transform is thus zero. However similar ensemble averages carried out on the autocorrelation function or energy spectrum do not average to zero so that an ensemble average has a useful meaning. This discussion based on ensemble averages (as opposed to time averages) illustrates a powerful technique for handling random signals which are not necessarily stationary random ones. Space does not allow further discussion here, except to remark that for stationary random functions the ensemble average technique gives results which are formally the same as those obtained using time averaging.

It is useful to recognise the significance of various time scales associated with stationary random functions. When examined over a very short interval of time such a function will appear smooth and of uniform slope $f'(t)$ (in the absence of delta functions and step discontinuities), and this will be so for intervals up to some characteristic limit t_A. As we examine the function over longer periods of time the characteristically random variations will be seen until, over intervals in excess of some duration time t_B, the statistical features will have established themselves and averages taken over any such interval will be approximately the same. The time t_B is approximately equal to the value of T in eqns (6.1)–(6.4) for which the expressions approximate to their asymptotic values. $R_f(t)$ is approximately constant for $t \gtrsim t_B$ and at the other extreme for $t \lesssim t_A$ then $R_f(t)$ is approximately equal to $R_f(0)$. In the two extremes we have

$$R_f(0) = \langle |f(t)|^2 \rangle \tag{6.11}$$

$$R_f(\infty) = |\langle f(t) \rangle|^2. \tag{6.12}$$

Expression (6.11) may be obtained by putting $t = 0$ in eqn (6.3) and expression (6.12) may be derived by writing $f(t)$ as the sum of its average value and a function representing the deviation of $f(t)$ from the average. The following expressions for real $f(t)$ are sometimes useful, involving the mean square change in value of $f(t_1)$ over a period t which we write as $\langle \Delta_f^2(t) \rangle$.

$$R_f(t) = \langle f^2(t) \rangle - \tfrac{1}{2} \langle \Delta_f^2(t) \rangle \tag{6.13}$$

$$\approx \langle f^2(t) \rangle - \frac{t^2}{2} \left\langle \left\{ \frac{df}{dt} \right\}^2 \right\rangle. \tag{6.14}$$

We prove these in Appendix K. Equation (6.13) shows clearly the transition between the two forms given in eqns (6.11) and (6.12) and establishes that $R_f(t)$ has its maximum value at $t = 0$.

Although a stationary random signal is an abstraction in that no experimental signal goes on for ever, it nevertheless represents a very good model for a signal whose duration is very long compared with the characteristic time t_B referred to above. The basic and most used statistical relation is certainly that expressed in eqn (6.8), since in so many situations the response of a system to a stimulus is determined by the power spectrum of the input signal. In the following sections we consider the form of the autocorrelation function and power spectrum for various special types of stationary random signal.

6.3   Signals with Non-Zero Mean

A stationary random signal whose mean value f_0 is non-zero has in its power spectrum a delta function of strength $2\pi f_0 f_0{}^*$ positioned at $\omega = 0$ superimposed on a distributed power spectrum characteristic of the fluctuations about the mean. To show this we split $f(t)$ into two parts and write it as

$$f(t) = f_0 + g(t)$$

then, bearing in mind that $g(t)$ has zero mean value:

$$R_f(t) = \lim_{T \to \infty} \frac{1}{2T} \int_{-T}^{+T} \{f_0{}^* + g^*(t_1)\} \{f_0 + g(t_1 + t)\} \, dt_1 \qquad (6.15)$$

$$= f_0 f_0{}^* + R_g(t). \qquad (6.16)$$

On Fourier transforming we get:

$$P_f(\omega) = 2\pi f_0 f_0{}^* \, \delta(\omega) + P_g(\omega). \qquad (6.17)$$

It is often convenient to substract off the mean value of a function during manipulation knowing that its effect can be reintroduced as above.

6.4   Sum of Signals

If $f(t), g(t)$, and $h(t)$ are finite power signals such that

$$f(t) = g(t) + h(t) \qquad (6.18)$$

then substitution of eqn (6.18) into the definition of $R_f(t)$ (eqn (5.18)) with subsequent use of the definition of cross-correlation (eqn (5.7)) soon leads to the following expression for the autocorrelation of $f(t)$:

$$R_f(t) = R_g(t) + R_h(t) + R_{gh}(t) + R_{hg}(t). \qquad (6.19)$$

Thus if $g(t)$ and $h(t)$ are uncorrelated, so that the cross-correlations in eqn (6.19) vanish, then the autocorrelations simply add. This will be so if $g(t)$ and $h(t)$ are uncorrelated random functions, and also if one of them is stationary random whilst the other is a periodic function. This fact provides the basis for a method of measuring a periodic signal which is 'buried' in noise, since the autocorrelation function consists of a periodic part (from the periodic signal) superimposed on a smooth aperiodic part arising from the noise (see Section 10.5).

6.5 Shot Noise

Shot noise is a special kind of random function produced by the successive repetition at random intervals of some square integrable function $h(t)$. The resultant function $f(t)$ is thus

$$f(t) = \sum_{-\infty}^{+\infty} h(t - t_i) \tag{6.20}$$

where the times t_i form a random sequence, the probability of an 'event' per unit time remaining constant. A typical situation in which such a signal arises is in electronic circuits when $h(t)$ represents the voltage or current fluctuation due to the transport of a single electron across some element and $f(t)$ is the overall resultant due to a random succession of electrons. A spatial analogue will occur in diffraction theory with a randomly arranged set of scattering centres: indeed when discussing the atomic arrangement in liquids, in Chapters 14 and 15, we shall see how to generalise the treatment given in this chapter in order to allow for the effects of correlation between the positions (or times) of different events.

We find that the power spectrum and autocorrelation function of $f(t)$ are directly related to the energy spectrum and autocorrelation function of $h(t)$. If on average there are v repetitions per unit time then:

$$P_f(\omega) = v\, S_h(\omega) + 2\pi\, |\langle f(t)\rangle|^2\, \delta(\omega) \tag{6.21}$$

$$R_f(t) = v\rho_h(t) + |\langle f(t)\rangle|^2. \tag{6.22}$$

The delta function in eqn 6.21 only exists if $f(t)$ has a non-zero average, and in electrical contexts corresponds to the D.C. component. As is fairly obvious the mean value of $f(t)$ is related to $h(t)$ by

$$\langle f(t)\rangle = v\int_{-\infty}^{+\infty} h(t)\, dt. \tag{6.23}$$

The above results are established in Appendix L.

A special case of shot noise is that of a random succession of delta functions. Putting $h(t) = \delta(t)$ in eqns (6.21) and (6.22) we get

$$P_f(\omega) = v + 2\pi v^2 \, \delta(\omega) \qquad (6.24)$$

$$R_f(t) = v\delta(t) + v^2. \qquad (6.25)$$

The power spectrum is evenly distributed over all frequencies, and the noise is in this case an example of 'white' noise.

Finally it may be useful for reference to quote without proof some statistical properties of shot noise which are often useful, but which are not directly related to Fourier transformation. With an average rate v the probability of an event in the interval t to $t + dt$ is $v\,dt$, whilst the probability of a time interval between successive events in the range t to $dt + dt$ is $ve^{-vt}\,dt$. The mean square deviation, σ^2, from the mean, f_0, is given by

$$\sigma^2 = \langle |f(t) - f_0|^2 \rangle = v \int_{-\infty}^{+\infty} |h(t)|^2 \, dt. \qquad (6.26)$$

For a real function the shot current tends towards a form possessing a normal distribution as $v \to \infty$, by which we mean that the probability of $f(t)$ having a value in the range f to $f + df$ at arbitrary time tends towards the value

$$\frac{1}{\sqrt{2\pi\sigma^2}} \exp\left\{ - (f - f_0)^2 / 2\,\sigma^2 \right\} df. \qquad (6.27)$$

The conditions under which these expressions hold are discussed more fully in books on random processes, for instance, reference 23, the condition for the validity of eqn (6.27) being approximately stated in the requirement that there must be considerable overlap of successive versions of $h(t)$ so that the duration of the signal $h(t)$ should be much greater than $1/v$.

6.6   Random Amplitude Modulation of a Carrier Wave

We consider first the complex function

$$f(t) = g(t) \, e^{i\omega_0 t} \qquad (6.28)$$

where $g(t)$ is a stationary random function. It is readily established by writing the autocorrelation functions out as integrals that

$$R_f(t) = e^{i\omega_0 t} R_g(t) \qquad (6.29)$$

and thence that

$$P_f(\omega) = P_g(\omega - \omega_0). \tag{6.30}$$

The effect of the random amplitude modulation is thus to broaden out the delta function power spectrum appropriate to the pure carrier wave.

Let us now consider the real counterpart of the above function, namely

$$f(t) = g(t) \cos \omega_0 t. \tag{6.31}$$

Somewhat surprisingly a general result analogous to the above eqn (6.30) is not available except in the limiting cases when the characteristic time τ of the random fluctuations is either much greater or much less than the period of the cosine term. In these extreme cases we get:

$$P_f(\omega) = \tfrac{1}{4}\{P_g(\omega - \omega_0) + P_g(\omega + \omega_0)\} \quad [\tau \gg 1/\omega_0] \tag{6.32}$$

$$P_f(\omega) = P_g(\omega) \quad [\tau \ll 1/\omega_0]. \tag{6.33}$$

A way of establishing these results is to consider the functions $f_T(t)$ and $g_T(t)$ obtained by truncating $f(t)$ and $g(t)$, so that

$$\left.\begin{aligned} f_T(t) &= f(t) \quad [-T < t < +T] \\ f_T(t) &= 0 \quad \text{otherwise} \end{aligned}\right\} \tag{6.34}$$

$$\left.\begin{aligned} g_T(t) &= g(t) \quad [-T < t < +T] \\ g_T(t) &= 0 \quad \text{otherwise} \end{aligned}\right\} \tag{6.35}$$

Letting $F_T(\omega)$ and $G_T(\omega)$ be the Fourier transforms it is readily established that

$$F_T(\omega)F_T^*(\omega) = \tfrac{1}{4}\{|G_T(\omega + \omega_0)|^2 + |G_T(\omega - \omega_0)|^2$$
$$+ G_T(\omega + \omega_0)G_T^*(\omega - \omega_0) + G_T(\omega - \omega_0)G_T^*(\omega + \omega_0)\}. \tag{6.36}$$

The two final terms on the right do not admit of simplification except in the extremes when the function $G_T(\omega)$ has a width which is either very small or very great compared with the value of ω_0, the cases corresponding respectively to $\tau \gg 1/\omega_0$ and $\tau \ll 1/\omega_0$. We have of course assumed that T is large enough not to introduce significant broadening. These two extremes give:

$$G_T(\omega + \omega_0)\,G_T^*(\omega - \omega_0) = 0 \quad [\tau \gg 1/\omega_0] \tag{6.37}$$

$$G_T(\omega + \omega_0) \approx G_T(\omega - \omega_0) \approx G_T(\omega) \quad [\tau \ll 1/\omega_0]. \tag{6.38}$$

The results in eqns (6.32) and (6.33) now follow from eqns (6.36), (6.37) and (6.38) if limits are taken in the usual way.

Fourier inversion of eqns (6.32) and (6.33) leads to the following expressions for the autocorrelation:

$$R_f(t) = \tfrac{1}{2} \cos \omega_0 t \, R_g(t) \qquad [\tau \gg 1/\omega_0] \qquad (6.39)$$

$$R_f(t) = R_g(t) \qquad [\tau \ll 1/\omega_0]. \qquad (6.40)$$

6.7   Random Phase Modulation

Introduction

We consider the function

$$f(t) = \exp\left[i(\omega_0 t + \phi(t))\right] \qquad [\langle \phi \rangle = 0] \qquad (6.41)$$

where $\phi(t)$ is a stationary random function of time. Such a signal arises for instance when the path difference between the emitter and receiver is varying randomly with time. Another situation arises in diffraction theory when a wavefront travels through a medium with randomly varying refractive index, the emerging wave front having a randomly distributed phase over its surface; in this case the time variable is replaced by a two-dimensional space variable. Our aim is to relate the power spectrum of $f(t)$ to the statistical properties of $\phi(t)$. A number of different relationships are possible in different situations and for clarity we will simply describe the results in the following sections leaving derivations to Appendix M. One generalisation is worth making. For weak modulation, with $\phi(t) \ll 2\pi$, then the power spectrum consists of a delta function at the carrier frequency ω_0 with a broad spectrum superimposed on it. As the modulation becomes stronger so the delta function becomes weakened until for strong modulation when the fluctuations in phase are much greater than 2π the carrier wave delta function is negligible. In the diffraction analogy this transition corresponds to the change from weak to strong scattering with the gradual disappearance of the directly transmitted beam. Note that frequency modulation produces a function given by eqn (6.41) and so is included in our discussion. The frequency at time t is simply $\omega_0 + d\phi/dt$.

The complex function in eqn (6.41) above is simpler to handle than its real counterpart $h(t)$ defined by

$$h(t) = \cos\{\omega_0 t + \phi(t)\}. \qquad (6.42)$$

However following a discussion similar to that in Section (6.6) it is not difficult to show that in the limits of 'slow' or 'fast' random fluctuations the power spectra of $f(t)$ and $h(t)$ are related as below. τ represents the characteristic time of the random fluctuations.

$$P_h(\omega) = \tfrac{1}{4}\{P_f(\omega) + P_f(-\omega)\} \quad [\tau \gg 1/\omega_0] \tag{6.43}$$

$$P_h(\omega) = P_f(\omega) \quad\quad\quad\quad [\tau \ll 1/\omega_0]. \tag{6.44}$$

General expression for the autocorrelation function

If $P(\phi, t)\,\mathrm{d}\phi$ is the probability that the phase will have changed by an amount in the range ϕ to $\phi + \mathrm{d}\phi$ in time t, then

$$R_f(t) = \mathrm{e}^{i\omega_0 t} \int \mathrm{e}^{i\phi} P(\phi, t)\,\mathrm{d}\phi. \tag{6.45}$$

The limits on the integration will be $-\infty$ to $+\infty$ or 0 to 2π depending on the convention chosen for describing phase changes. The proof is in Appendix M.

Special case with $\phi(t)$ normally distributed

If $\phi(t)$ is normally distributed then in eqn (6.45) we may substitute the following form for $P(\phi, t)$:

$$P(\phi, t) = \{2\pi\phi_0^2(t)\}^{-\frac{1}{2}} \exp\{-\phi^2/2\phi_0^2(t)\} \tag{6.46}$$

where $\phi_0^2(t)$ is the mean square change of phase in time t. It follows that

$$R_f(t) = \mathrm{e}^{i\omega_0 t} \exp\{-\phi_0^2(t)/2\} \tag{6.47}$$

$$= \mathrm{e}^{i\omega_0 t} \exp\{-\langle\{\phi(t)\}^2\rangle + R_\phi(t)\}. \tag{6.48}$$

Equation (6.48) follows from (6.47) by applying eqn (6.13) to $\phi(t)$.

Strong phase modulation with $\phi(t)$ normally distributed

If the modulation is 'strong' so that $\langle\phi^2(t)\rangle \gg 2\pi$ then a little thought applied to eqn (6.48) shows that the characteristic time of $f(t)$ will be much less than the characteristic time of $\phi(t)$. It is now possible to apply the short time approximation, eqn (6.14), to $\phi(t)$ so that eqn (6.48) becomes:

$$R_f(t) = \mathrm{e}^{i\omega_0 t} \exp\{-t^2 \langle(\mathrm{d}\phi/\mathrm{d}t)^2\rangle /2\}. \tag{6.49}$$

After Fourier transformation this gives

$$P_f(\omega) = \left\{\frac{2\pi}{\langle(\mathrm{d}\phi/\mathrm{d}t)^2\rangle}\right\}^{\frac{1}{2}} \exp\left\{\frac{-(\omega - \omega_0)^2}{2\langle(\mathrm{d}\phi/\mathrm{d}t)^2\rangle}\right\}. \tag{6.50}$$

The power spectrum of $f(t)$ is thus Gaussian irrespective of the actual form of the power spectrum of $\phi(t)$. This expression is particularly convenient if the phase modulation results from frequency modulation since the frequency at time t is $(\omega_0 + d\phi/dt)$ and eqn (6.50) thus gives $P_f(\omega)$ in terms of the mean square deviation of the frequency.

Weak phase modulation

For the case that $\langle \phi^2(t) \rangle \ll 2\pi$ we have:

$$R_f(t) = e^{i\omega_0 t} \{1 - \langle \phi^2(t) \rangle + R_\phi(t)\} \qquad (6.51)$$

$$P_f(\omega) = 2\pi \{1 - \langle \phi^2(t) \rangle\} \delta(\omega - \omega_0) + P_\phi(\omega - \omega_0). \qquad (6.52)$$

This can be derived as a limiting form of eqn (6.48), but as shown in Appendix M the result does not require the assumption of a normal distribution.

Sudden random changes of phase

Suppose that the function $\phi(t)$ suffers sudden changes in value at random time intervals, the new value of $\phi(t)$ after each change having a uniform probability distribution within the range of values 0 to 2π. In between jumps $\phi(t)$ remains constant. If on average there are v changes per unit time, then the probability of no change having occurred during a time interval t is $e^{-v|t|}$. Into eqn (6.45) we may substitute the expression

$$P(\phi, t) = e^{-v|t|} \delta(\phi) + \frac{1}{2\pi} (1 - e^{-v|t|}) \qquad (6.53)$$

In this expression the first term on the right hand side represents the case that ϕ has not changed during time t, and the other term represents the remaining probability distributed uniformly over the range 2π. Eqn (6.45) now gives,

$$R_f(t) = e^{i\omega_0 t} e^{-v|t|} \qquad (6.54)$$

and Fourier transformation gives

$$P_f(\omega) = \frac{2v}{v^2 + (\omega - \omega_0)^2} . \qquad (6.55)$$

6.8   Random Telegraph Signals—First Type

We consider the function to have fixed magnitude A, but to change sign at a succession of randomly chosen instants, the probability of remaining a time t before changing sign being ve^{-vt} where v is the mean number of sign changes per unit time. We find, as shown in Appendix N.

$$R_f(t) = A^2 e^{-2\nu|t|} \tag{6.56}$$

$$P_f(\omega) = \frac{4A^2\nu}{4\nu^2 + \omega^2}. \tag{6.57}$$

6.9   Random Telegraph Signal—Second Type

As in Section (6.8) we consider the function $f(t)$ to have constant magnitude A, but now consider that at regularly spaced instants, a time τ apart, the sign of $f(t)$ is reassigned at random, the chances of a positive or negative sign being equal at each reassignment. We find (as shown also in Appendix N):

$$R_f(t) = A^2\left(1 - \frac{|t|}{\tau}\right) \quad [|t| \leqslant \tau] \tag{6.58}$$

$$= 0 \qquad\qquad [|t| > \tau] \tag{6.59}$$

$$P_f(\omega) = A^2\tau \left\{ \frac{\sin(\omega\tau/2)}{(\omega\tau/2)} \right\}^2. \tag{6.60}$$

PART II—APPLICATIONS

Chapter 7

Linear Systems

7.1 Introduction

We describe here how Fourier transform techniques are useful in computing the response of a rather general class of systems to an arbitrary 'driving force' or input. By system we mean simply a 'black box' to which we apply an input $f(t)$ and from which comes a response $g(t)$. For instance with a mechanical oscillator we apply a force $f(t)$ and the response is a displacement $g(t)$; in an electrical system we apply a current or voltage $f(t)$ and the response is also a current or voltage; in diffraction problems discussed in Chapter 11 we 'apply' a beam of radiation with an angular spectrum $f(\mathbf{q})$ and the response is a diffracted beam with a different angular spectrum $g(\mathbf{q})$. Although as this last example shows t and ω are not the only pairs of variables in which our concepts are useful, we will in this chapter base our discussion on this pair of variables on account of the extensive use of the concept in electrical engineering.

One method of finding the response of a mechanical or electrical system to an input is to solve in detail the differential equations governing the system; this will be illustrated in Chapter 8 for the damped harmonic oscillator. Another approach relies not on the detailed 'equations of motion' but simply on a knowledge of the response of the system to an applied impulse of unit strength. We call this response the impulse response function and will denote it by $h(t)$, so that in other words when $f(t) = \delta(t)$ then $g(t) = h(t)$. A knowledge of this impulse response provides all the information necessary to calculate the output for any input signal, provided the system satisfies certain conditions referred to below. Of course although computationally a knowledge of the impulse response provides all one needs to know about the system in order to calculate an output corresponding to a given input, it does not provide an understanding of the system in the same way that 'opening' the black box and 'looking inside' does. It is important also to remember that the method is only straightforwardly applicable to 'linear shift invariant systems'. A linear system is one such that if signals $f_1(t)$ and $f_2(t)$ give outputs $g_1(t)$ and $g_2(t)$, then an input $\{f_1(t) + f_2(t)\}$ gives an output $\{g_1(t) + g_2(t)\}$. A system is shift invariant if delaying an input has no effect other than to delay the output by the same amount. We shall prove that if a harmonic signal is applied to a linear shift invariant system then the output consists simply of a harmonic function at the same frequency. Thus one test of a linear system is to apply a harmonic signal and to analyse the output for components at other frequencies: the presence of such components will indicate that the system is not linear. Another test of course is to see whether doubling the input doubles the output: if not then once again the system must be non-linear. Thus although a large class of active and passive electrical circuits are linear, the presence of saturation or hysteresis in transformers or 'rounded' characteristics in valves or transistors can destroy the validity of this approach.

7.2 Impulse Response Function and Transfer Function

We have already defined the impulse response $h(t)$ as the output for an input $\delta(t)$, and we now define the transfer function $H(\omega)$ as the Fourier transform of $h(t)$:

$$H(\omega) = \int_{-\infty}^{+\infty} e^{-i\omega t} h(t) \, dt. \tag{7.1}$$

We are now ready to quote the result which is central to our development of linear systems, namely that the input $f(t)$ and output $g(t)$ of a linear shift invariant system are related by:

$$g(t) = f(t) \otimes h(t) \tag{7.2}$$

$$G(\omega) = F(\omega) H(\omega) \tag{7.3}$$

where

$$G(\omega) = \int_{-\infty}^{+\infty} e^{-i\omega t} g(t) \, dt \tag{7.4}$$

and

$$F(\omega) = \int_{-\infty}^{+\infty} e^{-i\omega t} f(t) \, dt. \tag{7.5}$$

Equations (7.2) and (7.3) follow from each other through the convolution theorem (Section 5.4) so that it is sufficient to establish either of them as a starting point. Equation (7.2) has already been arrived at in a rather intuitive way in Section 5.1, the argument being that if the system 'spreads' a delta function so as to become $h(t)$ then it spreads a function of finite duration so as to become $f(t) \otimes h(t)$. The argument is made more rigorous in Appendix O, the crux of the proof lying in our being able to consider a continuous function $f(t)$ as equivalent to a series of closely spaced delta functions of appropriate strengths. Equation 7.3 is also plausible in its own right if we accept that our discussion is limited to systems for which a harmonic input gives a harmonic output at the same frequency, but altered perhaps in amplitude and phase. We now interpret $F(\omega)$ and $G(\omega)$ as the complex amplitudes of harmonic input and outputs, and regard $H(\omega)$ as a complex transfer function telling us the change in amplitude and phase of the wave as it passes through the system.

It is clear that eqn (7.3) will apply to any signals that are Fourier transformable including ones that involve delta functions in their transform, provided the product $F(\omega) H(\omega)$ does not involve a product of delta functions at the same frequency. Thus stationary random functions are excluded (though a finite portion of such a signal is allowed: see Section 7.5) and in practice eqn (7.3) is applicable when one or other of the functions $f(t)$ and $h(t)$ is finite energy, and the other is either finite energy or periodic. When time is the variable $h(t)$ possesses the further property of being 'casual' by which we mean that $h(t)$ is zero for $t < 0$. This is just an expression of the causality of nature in that the 'effect' must follow the cause. However such a restriction on $h(t)$ is not necessary for the validity of the equations which we develop in this chapter, and it should be remembered that when we apply our results to optical systems in which a space variable replaces the time variable, then we frequently encounter impulse responses which are not casual: as we shall see in Chapter 11 the corresponding function is then often called the 'point spread' function. When a signal $f(t)$ has been recorded as a function of time it can be processed by a computer to give a simulated 'output' $g(t)$ and here again there is no need for the impulse response to be casual since the time variable is not in this case 'real time'. We discuss the implications of causality further in Section 9.6 and in Appendix U.

Note that even if $f(t), g(t)$ and $h(t)$ are real quantities the transfer function $H(w)$ is an essentially complex quantity. A real input can however be represented by the real (or imaginary) part of a complex function $f(t)$, and due to the linearity of the system the output will be the real (or imaginary) part of the function $g(t)$ calculated through eqns (7.2) and (7.3). Thus complex numbers arise in two possible ways and confusion can arise if this is not realized.

7.3   Harmonic Input. Step Input and Step Response

One way of finding the impulse response of a system is to apply the nearest approximation one can to an impulse and to measure the response. Clearly however this is not always convenient, and there are alternative ways based on applying either a harmonic input or a step function input. Let us consider these in turn.

Application of a harmonic input gives a harmonic output, as indicated in the three equations below with the arrows pointing from input to output.

$$e^{i\omega_0 t} \to H(\omega_0)e^{i\omega_0 t} \tag{7.6}$$

$$\cos \omega_0 t \to A(\omega_0) \cos \{\omega_0 t + \phi(\omega_0)\} \tag{7.7}$$

$$\sin \omega_0 t \to A(\omega_0) \sin \{\omega_0 t + \phi(\omega_0)\} \tag{7.8}$$

where $A(\omega)$ and $\phi(\omega)$ are real quantities defined from $H(\omega)$ by

$$H(\omega) = A(\omega)e^{i\phi(\omega)}. \tag{7.9}$$

Equation (7.6) may be readily obtained by using eqn (7.3) to derive:

$$G(\omega) = 2\pi H(\omega)\,\delta(\omega - \omega_0) \qquad [\text{for } f(t) = e^{i\omega_0 t}]$$

and then inversely transforming to obtain the output $g(t)$. Equations (7.7) and (7.8) follow as the real and imaginary parts of eqn (7.6). Thus if sine or cosine functions of various frequencies are applied to a system and $A(\omega)$ and $\phi(\omega)$ derived from the amplitudes and phase advances of the outputs, then $H(\omega)$ is known directly from eqn (7.9). For the special case of an alternating current being the input, and the voltage developed across some element being the output we see that $H(\omega)$ plays a role very similar to that of the impedance of the element: if we apply an alternating voltage as input and measure the current flow as the output then $H(\omega)$ is similar to the complex admittance.

It is interesting and instructive to see that the linear and shift invariant properties of a system lead in a very direct way to the requirement that a harmonic input requires a harmonic output at the same frequency. The argument runs as follows:

if

$$e^{i\omega_0 t} \to g(t) \tag{7.10}$$

then linearity requires that

$$e^{i\omega_0 t_0} e^{i\omega_0 t} \to e^{i\omega_0 t_0} g(t) \tag{7.11}$$

and shift invariance requires that

$$e^{i\omega_0(t+t_0)} \rightarrow g(t + t_0).\tag{7.12}$$

However eqns (7.11) and (7.12) must be equivalent so that

$$e^{i\omega_0 t_0} g(t) = g(t + t_0).\tag{7.13}$$

This result can be satisfied only if $g(t)$ is proportional to $e^{i\omega_0 t}$, as required.
Let us consider now the effect of applying a step input $U(t)$ such that

$$\left.\begin{array}{ll} U(t) = 1 & [t > 0] \\ = 0 & [t < 0] \end{array}\right\}.\tag{7.14}$$

Let us define the resulting output $b(t)$ as the step response and see how $b(t)$ and its transform $B(\omega)$ defined by

$$B(\omega) = \int e^{-i\omega t} b(t)\, \mathrm{d}t\tag{7.15}$$

are related to $h(t)$ and $H(\omega)$. Equation (7.2) allows us to write

$$b(t) = h(t) \otimes U(t)$$

$$= \int_{-\infty}^{+\infty} h(t_1)\, U(t - t_1)\, \mathrm{d}t_1$$

$$= \int_{-\infty}^{t} h(t_1)\, \mathrm{d}t_1.\tag{7.16}$$

Clearly it follows from eqn (7.16) that

$$h(t) = \frac{\mathrm{d}}{\mathrm{d}t} b(t).\tag{7.17}$$

This result eqn (7.17) is satisfactorily in agreement with what we might expect if we regard the delta function input as the derivative of a step input. To relate $H(\omega)$ and $B(\omega)$ together we utilise eqn (7.3), using eqn (2.77) to give us the transform of $U(t)$.

$$B(\omega) = H(\omega) \left\{ \pi\delta(\omega) - \frac{i}{\omega} \right\}\tag{7.18}$$

$$= \pi H(0)\delta(\omega) - \frac{iH(\omega)}{\omega}\tag{7.19}$$

so that $$H(\omega) = i\omega B(\omega) \tag{7.20}$$

We have assumed in eqns (7.18) to (7.20) that $H(\omega)$ contains no delta function at $\omega = 0$; this is a physically safe assumption it being easily verified that such a delta function would correspond to a step response $b(t)$ that diverged with time. In the absence of singularities eqn (7.20) may alternatively be derived from (7.17) using the differentation theorem eqn (2.29).

A close connection between the step response and the impulse response is of course to be expected if an impulse is regarded as the result of differentiating a step function. For instance applying an impulse of current to a capacitor is equivalent physically to applying a step function of charge so that the choice of treatment in this case depends simply upon whether charge or current is used as the variable.

7.4   Energy Transfer Function

If we define $|F(\omega)|^2$ and $|G(\omega)|^2$ as the energy spectra of the input and output signals, then clearly from eqn (7.3)

$$|G(\omega)|^2 = |F(\omega)|^2 |H(\omega)|^2 \tag{7.21}$$

and $|H(\omega)|^2$ is called the energy transfer function of the system. If the rate of energy dissipation at the output is instantaneously proportional to $|g(t)|^2$ then the total energy E dissipated in all time may be written as follows using Parseval's theorem (eqn (2.27)) and using R to represent a resistive constant when $g(t)$ represents a current or velocity.

$$E = R \int_{-\infty}^{+\infty} |g(t)|^2 \, dt \tag{7.22}$$

$$= \frac{R}{2\pi} \int_{-\infty}^{+\infty} |G(\omega)|^2 \, d\omega \tag{7.23}$$

$$= \frac{R}{2\pi} \int_{-\infty}^{+\infty} |F(\omega)|^2 \, |H(\omega)|^2 \, d\omega \tag{7.24}$$

If $f(t)$, $g(t)$ and $h(t)$ are real functions then $|F(\omega)|^2$ and $|H(\omega)|^2$ are even functions of ω and the above simplifies to

$$E = \frac{R}{\pi} \int_{0}^{\infty} |F(\omega)|^2 |H(\omega)|^2 \, d\omega . \tag{7.25}$$

It is these expressions (7.24) and (7.25) which justify the name 'energy spectrum' for quantities like $|F(\omega)|^2$ since they confirm the idea that energy is presented to the system over a range of frequencies and that the system selectively accepts certain regions of this energy spectrum in a way determined by its own characteristic function $|H(\omega)|^2$. The analogy with polychromatic light passing through a filter is of course exact.

Treatments analogous to the above may be developed to cover slightly different situations. For instance the rate of dissipation may be proportional to $\{\dot{g}(t)\}^2$ as is the case when $g(t)$ represents a displacement and viscous damping is present. Using the differentiation theorem in conjunction with Parseval's theorem (see eqns (E.9–E.11)) we soon derive (for $\dot{g}(t)$ real)

$$E = (2\lambda_0) \int_{-\infty}^{+\infty} \{\dot{g}(t)\}^2 \, dt \tag{7.26}$$

$$= \frac{2\lambda_0}{\pi} \int_0^\infty \omega^2 \, |F(\omega)|^2 \, |H(\omega)|^2 \, d\omega. \tag{7.27}$$

The proportionality constant has been chosen as $2\lambda_0$ so as to follow the notation to be adopted in Chapter 8. In electromagnetic theory the rate of radiation of energy from a moving charge is given (for speeds small compared with the velocity of light) by:

$$\frac{dE}{dt} = -\frac{2}{3} \frac{e^2}{4\pi\varepsilon_0 c^3} \dddot{g}(t) \dot{g}(t) \tag{7.28}$$

where e, ε_0 and c are the charge, permittivity of free space and velocity of light in MKS units, and $g(t)$ is the real displacement of the charge. In this case we derive for the energy radiated

$$E = -\frac{2}{3} \frac{e^2}{4\pi\varepsilon_0 c^3} \int_{-\infty}^{+\infty} \dddot{g}(t) \dot{g}(t) \, dt \tag{7.29}$$

$$= +\frac{2}{3} \frac{e^2}{4\pi\varepsilon_0 c^3} \frac{1}{\pi} \int_0^\infty \omega^4 \, |F(\omega)|^2 \, |H(\omega)|^2 \, d\omega. \tag{7.30}$$

This result is applicable in discussing the scattering of radiation from atoms, with the electrons regarded as oscillators driven by the applied fields.

The above expressions relate to the energy dissipated in all time at the output of a system. The total energy applied may also be treated. For instance if $f(t)$ represents an applied voltage and $g(t)$ represents the current produced then we may derive (for $f(t)$ and $g(t)$ real)

$$E = \int_{-\infty}^{+\infty} f(t) g(t) \, dt \tag{7.31}$$

$$= \frac{1}{2\pi} \int_{-\infty}^{+\infty} |F(\omega)|^2 \, H^*(\omega) \, d\omega = \frac{1}{\pi} \int_{0}^{\infty} [\mathrm{Re}\, H(\omega)] \, |F(\omega)|^2 \, d\omega. \tag{7.32}$$

Alternatively if $f(t)$ represents an applied force and $g(t)$ represents a resultant displacement then we derive (again for $f(t)$ and $g(t)$ real)

$$E = \int_{-\infty}^{+\infty} f(t) \, \dot{g}(t) \, dt \tag{7.33}$$

$$= \frac{1}{2\pi} \int_{-\infty}^{+\infty} i\omega \, |F(\omega)|^2 \, H(\omega) \, d\omega$$

$$= \frac{1}{\pi} \int_{0}^{\infty} \omega \, [- \mathrm{Im}\, H(\omega)] \, |F(\omega)|^2 \, d\omega. \tag{7.34}$$

7.5   Power Transfer. Random or Periodic Input

The results of Section (7.4) apply only to finite energy signals at input and output. However we may adapt the results to cover finite power signals such as periodic or stationary random signals quite easily. We first truncate the input and output so as to create finite energy signals $f_T(t)$ and $g_T(t)$ with the corresponding transforms $F_T(\omega)$ and $G_T(\omega)$. These functions are equal respectively to $f(t)$ and $g(t)$ for $|t| \leqslant T$ but are zero otherwise. Clearly the energy transfer E_T due to these truncated signals is given by the expressions in Section 7.4, and we merely have to divide through by $2T$ and take the limit as $T \to \infty$ to obtain expressions for the mean rate of energy transfer $\langle W \rangle$. Thus if we define

$$\langle W \rangle = \lim_{T \to \infty} \frac{E_T}{2T} \tag{7.35}$$

$$P_f(\omega) = \lim_{T \to \infty} \frac{1}{2T} |F_T(\omega)|^2 \tag{7.36}$$

$$P_g(\omega) = \lim_{T \to \infty} \frac{1}{2T} |G(\omega)|^2 \tag{7.37}$$

then we may for instance rewrite eqns (7.21) and (7.25) as

$$P_g(\omega) = P_f(\omega) \, |H(\omega)|^2 \qquad (7.38)$$

$$\langle W \rangle = \frac{R}{\pi} \int_0^\infty P_f(\omega) \, |H(\omega)|^2 \, d\omega \qquad (7.39)$$

whilst (7.27) and (7.34) become:

$$\langle W \rangle = \frac{2\lambda_0}{\pi} \int_0^\infty \omega^2 \, P_f(\omega) \, |H(\omega)|^2 \, d\omega \qquad (7.40)$$

and

$$\langle W \rangle = \frac{1}{\pi} \int_0^\infty - \omega \, P_f(\omega) \, [\mathrm{Im} \, H(\omega)] \, d\omega. \qquad (7.41)$$

Chapter 8

Response of a Damped Harmonic Oscillator to a Driving Force

8.1 Introduction

The damped harmonic oscillator provides a prototype for resonance absorption and scattering in acoustic, electrical, atomic and nuclear systems. Treatment of the problem shows clearly the two different approaches that are possible, one method using Fourier techniques being based on the impulse response and transfer function (see Chapter 7), the other method being based on detailed solution of the differential equation governing the system. The general problem is to solve the equation

$$\ddot{x} + 2\lambda_0 \dot{x} + (\omega_0{}^2 + \lambda_0{}^2)\, x = f(t). \tag{8.1}$$

For a mechanical oscillator $f(t)$ is the externally applied driving force per unit mass, whilst the terms in $\dot{x}$ and x represent respectively a damping force and a restoring force per unit mass. In electric circuit theory this equation governs the behaviour of a series resonant circuit with inductance L, capacitance C, and resistance R. When an E.M.F. $V(t)$ is applied we interpret the terms as follows:

$$f(t) = \frac{V(t)}{L}$$

$$2\lambda_0 = \frac{R}{L}$$

$$(\omega_0{}^2 + \lambda_0{}^2) = \frac{C}{L} \qquad (8.2)$$

$$x = \text{charge}$$

$$\dot{x} = \text{current}$$

The electromagnetic problem of the motion of a charge under the influence of an applied electric force is also determined by an eqn like (8.1) above. However in this case an extra term equal to

$$-\frac{2}{3} \frac{1}{4\pi\,\varepsilon_0} \frac{e^2}{mc^3} \dddot{x}$$

must be added on the left-hand side to allow for a force per unit mass which arises due to the radiation of energy (see Section 7.4). The effect of this extra term may be approximately allowed for simply by modifying the value of λ_0 in eqn (8.1) by the addition of a part proportional to ω^2. Such a treatment underlies the classical theories of light and X-ray scattering by atoms and nuclei.

8.2   Response to Harmonic and Impulse Inputs

A simple special case to consider is that of a complex harmonic driving force per unit mass, so that we have to solve eqn (8.1) with

$$f(t) = Ae^{i\omega_1 t}.$$

Substitution of a trial solution of the form $x(t) = H(\omega_1)Ae^{i\omega_1 t}$ soon establishes that this is indeed a solution provided the function H is of the form:

$$H(\omega) = \frac{1}{(\lambda_0{}^2 + \omega_0{}^2 - \omega^2) + 2i\lambda_0\omega}. \qquad (8.3)$$

Comparison with eqn (7.6) shows that (8.3) gives us the transfer function of the system with force per unit mass regarded as input and displacement as output.

Another simple case to consider is the response to an impulsive force. This has the effect of giving the particle an initial velocity after which it moves under the influence of zero external force so that we solve eqn (8.1) with $f(t) = 0$. Direct substitution into eqn (8.1) soon verifies that a solution of the following form exists:

$$x(t) \propto \sin \omega_0 t \, e^{-\lambda_0 t} \qquad [t > 0]. \tag{8.4}$$

If the impulsive force at $t = 0$ is regarded as the limiting case of a steady large force acting for a short time (as the duration tends to zero and the magnitude tends to infinity) then it is easily verified that such an impulse produces unit velocity. Using this as an initial condition determining the constant of proportionality in eqn (8.4) we arrive finally at the impulse response, $h(t)$, which comes to be:

$$h(t) = \frac{1}{\omega_0} \sin \omega_0 t \, e^{-\lambda_0 t} \qquad [t > 0]. \tag{8.5}$$

Now according to the definition of transfer function adopted in eqn (7.1) our functions $H(\omega)$ and $h(t)$ given in eqns (8.3) and (8.5) should be a Fourier transform pair, and reference to eqn (2.47) confirms that this is so. Thus we could have derived eqn (8.5) form (8.3) or vice versa.

By considering the real part of the solution given above for a complex harmonic input we find that an input

$$f(t) = A \cos \omega_1 t \tag{8.6}$$

gives a response

$$x(t) = x_0 \cos (\omega_1 t + \phi) \tag{8.7}$$

where

$$\tan \phi = \frac{-2\omega_1 \lambda_0}{(\omega_0{}^2 - \omega_1{}^2 + \lambda_0{}^2)} \quad [-\pi < \phi < 0] \tag{8.8}$$

and

$$x_0 = A\{(\lambda_0{}^2 + \omega_0{}^2 - \omega_1{}^2)^2 + 4\lambda_0{}^2\omega_1{}^2\}^{-\frac{1}{2}} \qquad [\text{Exact}] \tag{8.9}$$

$$\approx (A/2\omega_0)/\{(\omega_0 - \omega_1)^2 + \lambda_0{}^2\}^{\frac{1}{2}} \, [\lambda_0 \ll \omega_0, |\omega_1 - \omega_0| \ll \omega_0] \tag{8.10}$$

$$\approx |A/(\omega_0{}^2 - \omega_1{}^2)| \quad [|\omega_1 - \omega_0| \gg \lambda_0] \tag{8.11}$$

$$\approx A/\omega_0{}^2 \qquad [\omega_1 \ll \omega_0, \lambda_0 \ll \omega_0] \tag{8.12}$$

$$\approx A/\omega_1{}^2 \qquad [\omega_1 \gg \omega_0, \omega_1 \gg \lambda_0]. \tag{8.13}$$

Note the Lorentzian form for the approximation at frequencies near a sharp resonance, eqn (8.10), and the other approximations (8.11)–(8.13) when the frequency is far from resonance.

We may now consider the energy dissipated per unit time. The instantaneous rate of energy dissipation per unit mass is $2\lambda_0(\dot{x})^2$ so that the mean rate of dissipation $\langle W \rangle$ is given by

$$\langle W \rangle = \langle 2\lambda_0(\dot{x})^2 \rangle \tag{8.14}$$

$$= \lambda_0 x_0^2 \omega_1^2 \tag{8.15}$$

$$= A^2 \lambda_0 \omega_1^2 \{(\lambda_0^2 + \omega_0^2 - \omega_1^2)^2 + 4\lambda_0^2 \omega_1^2\}^{-1} \tag{8.16}$$

$$\approx \frac{A^2 \lambda_0}{4} \frac{1}{(\omega_0 - \omega_1)^2 + \lambda_0^2} \qquad [\lambda_0 \ll \omega_0, |\omega_1 - \omega_0| \ll \omega_0]. \tag{8.17}$$

It is satisfactory that this result, eqn (8.16), can also be obtained from eqns (7.40) or (7.41) if we utilise the fact that for our cosine wave input we have (see eqns (2.59) and (4.11))

$$P_f(\omega) = \frac{\pi A^2}{2} \{\delta(\omega - \omega_1) + \delta(\omega + \omega_1)\}. \tag{8.18}$$

8.3   Damped Harmonic Driving Force

Now that we have a knowledge of the impulse response $h(t)$ and the transfer function $H(\omega)$ the techniques developed in Chapter 7 can be used to find the response and energy dissipation for an input of any type. Let us choose as an example the damped harmonic input:

$$f(t) = A \cos \omega_1 t \, e^{-\lambda_1 t} \qquad [t \geqslant 0]$$
$$= 0 \qquad [t < 0]. \tag{8.19}$$

The resulting differential eqn (8.1) can be solved by standard techniques to give:

$$x(t) = (a \cos \omega_0 t + b \sin \omega_0 t)e^{-\lambda_0 t} + (c \cos \omega_1 t + d \sin \omega_1 t)e^{-\lambda_1 t} \tag{8.20}$$

where

$$c = A\{\omega_0^2 - \omega_1^2 + (\lambda - \lambda_0)^2\}/[\{\omega_0^2 - \omega_1^2 + (\lambda_1 - \lambda_0)^2\}^2$$
$$+ 4\omega_1^2(\lambda_1 - \lambda_0)^2] \tag{8.21}$$

$$d = \frac{2A\omega_1(\lambda_0 - \lambda_1)}{\{\omega_0{}^2 - \omega_1{}^2 + (\lambda_1 - \lambda_0)^2\}^2 + 4\omega_1{}^2(\lambda_1 - \lambda_0)^2} \, . \tag{8.22}$$

The constants a and b are determined by the initial conditions and for the condition $x = 0, \dot{x} = 0$ at $t = 0$ we have:

$$a = -c \tag{8.23}$$

$$b = \{c(\lambda_1 - \lambda_0) - \omega_1 d\}/\omega_0. \tag{8.24}$$

The solution in eqn (8.20) exhibits many well-known facts, the amplitude of the oscillatory response rising from zero to a maximum and then dying away again. The solution may be regarded as the sum of two contributions the 'complementary function' and the 'particular integral' being the terms on the right-hand side of eqn (8.20). The complementary function represents a 'ringing' term giving oscillation at the natural period of the system, whilst the particular integral represents a motion following the applied force but out of phase. If the applied force is of short duration and the damping small then after a complicated initial motion the system settles down to an oscillatory motion of decreasing amplitude at the natural frequency (as when a bell is struck). In this case the complementary function dominates. On the other hand if a force of long duration is applied to a highly damped system then the system settles down to a motion oscillating at the frequency of the applied force.

The results just described can alternatively be described in the frequency domain since according to eqn (7.3) we have

$$X(\omega) = F(\omega)H(\omega) \tag{8.25}$$

where

$$X(\omega) = \int e^{-i\omega t} x(t) \, dt$$

and

$$F(\omega) = \int e^{-i\omega t} f(t) \, dt.$$

If for simplicity we consider the complex applied force

$$f(t) = Ae^{i\omega t_1} e^{-\lambda t_1} \quad [t \geqslant 0]$$

$$= 0 \qquad\qquad [t < 0] \tag{8.26}$$

then for an applied signal of long duration applied to a highly damped oscillator we find that $F(\omega)$ is sharply peaked about $\omega = \omega_1$, whilst $H(\omega)$ is a broad function. As a result we may write

$$X(\omega) \approx F(\omega) H(\omega_1) \tag{8.27}$$

showing that $x(t)$ is determined mainly by $f(t)$ since $H(\omega)_1$ is a constant. On the other hand for an applied signal of short duration applied to a lightly damped oscillator we find that $F(\omega)$ is a broad function whilst $H(\omega)$ is sharply peaked at $\omega = \omega_0$ and $\omega = -\omega_0$. As a result we write

$$X(\omega) \approx F(\omega_0)\,H(\omega) \qquad [\omega > 0] \tag{8.28}$$

$$\approx F(-\omega_0)\,H(\omega) \quad [\omega < 0]$$

showing that $x(t)$ is determined mainly by the properties of $h(t)$.

Let us now consider the total energy dissipated in all time. The rate of dissipation is $2\lambda_0(\dot{x})^2$ so that we could proceed by evaluating

$$E = \int_0^\infty 2\lambda_0(\dot{x})^2 \, dt \tag{8.29}$$

with $x(t)$ given by eqn (8.20). The working is clearly very arduous, and it is now very useful and instructive to work in the frequency domain using eqns (7.27) or (7.34). Eqns (7.27), (2.46) and (8.3) give us:

$$E = \frac{2\lambda_0}{\pi} \int_0^\infty \omega^2 \, |F(\omega)|^2 \, |H(\omega)|^2 \, d\omega \tag{8.30}$$

where

$$|F(\omega)|^2 = A^2 \left[\frac{\lambda_1^2(\lambda_1^2 + \omega^2 + \omega_1^2)^2 + \omega^2(\lambda_1^2 + \omega^2 - \omega_1^2)^2}{\{(\lambda_1^2 + \omega_1^2 - \omega^2)^2 + 4\lambda_1^2\omega^2\}^2} \right] \tag{8.31}$$

$$\approx \frac{A^2}{4} \frac{1}{(\omega - \omega_1)^2 + \lambda_1^2} \quad [\lambda_1 \ll \omega_1, |\omega_1 - \omega| \ll \omega_1] \tag{8.32}$$

and

$$|H(\omega)|^2 = \frac{1}{(\lambda_0^2 + \omega_0^2 - \omega^2)^2 + 4\lambda_0^2\omega^2} \tag{8.33}$$

$$\approx \frac{1/4\omega_0^2}{(\omega - \omega_0)^2 + \lambda_0^2} \quad [\lambda_0 \ll \omega_0, |\omega_0 - \omega| \ll \omega_0]. \tag{8.34}$$

The working is still tedious, but eqn (8.30) shows how the energy dissipated depends on the overlap of two spectra characteristic of the applied force and the system respectively. If we use the approximations appropriate for light damping then we get:

$$E \approx \frac{\lambda_0 A^2}{8\pi} \int_0^\infty \frac{\omega^2}{{\omega_0}^2} \frac{1}{(\omega - \omega_1)^2 + {\lambda_1}^2} \frac{1}{(\omega - \omega_0)^2 + {\lambda_0}^2} \, d\omega \qquad (8.35)$$

$$\approx \frac{A^2}{8\lambda_1} \frac{\lambda_0 + \lambda_1}{(\omega_0 - \omega_1)^2 + (\lambda_0 + \lambda_1)^2} \,. \qquad (8.36)$$

Note that eqn (8.35) represents the convolution of two Lorentzians together and eqn (8.36) follows as a standard result whose derivation is referred to in Section 5.4.

Chapter 9

Passive Electric Circuits Treated as Linear Systems

9.1 Introduction

In this chapter we exemplify the ideas introduced in Chapter 7 by applying them to various simple passive electric circuits. We first consider two terminal arrangements in which the voltage and current are regarded as the input and output. We then consider four terminal systems (two input terminals and two output terminals) as exemplified by various classic forms of filter network and by the transmission line. Following these discussions of actual systems we conclude with some discussion of filters in general introducing the concepts of causality and of the ideal filter.

9.2 Two Terminal *L, R, C* circuits

Let us first consider simply a capacitor, with a current $I(t)$ regarded as input, and a resulting voltage $V(t)$ regarded as output. Our aim is to relate $I(t)$ and $V(t)$ by finding the impulse response $h(t)$ and the transfer function $H(\omega)$, as defined in Sections 7.1 and 7.2.

An impulse of current, $I(t) = \delta(t)$, is equivalent to a sudden application of unit charge, which in turn results in the sudden appearance of a voltage equal to $1/C$ since the capacity C is defined as the ratio of charge to voltage. Thus we have an impulse response given by

$$
\begin{aligned}
h(t) &= 1/C \quad [t \geqslant 0] \\
&= 0 \quad\quad\ [t < 0]
\end{aligned}
\Bigg\}.
\tag{9.1}
$$

By Fourier transforming using eqn (7.1) and (2.77) we now obtain

$$H(\omega) = \frac{\pi}{C}\delta(\omega) - \frac{i}{\omega C} \tag{9.2}$$

$$= \frac{\pi}{C}\delta(\omega) + \frac{1}{\omega C}e^{-i\pi/2}. \tag{9.3}$$

Equations (7.2) and (7.3) now allow us to evaluate $V(t)$ from any $I(t)$, in principle.

It is interesting to note that for $\omega > 0$ the transfer function $H(\omega)$ is identical to the impedance $\mathbf{Z}(\omega)$, familiar from A.C. theory as

$$\mathbf{Z}(\omega) = \frac{-i}{\omega C}. \tag{9.4}$$

This is to be expected since according to eqns (7.6)–(7.9) a harmonic input gives rise to a harmonic output for any linear shift invariant system, with $H(\omega)$ playing precisely the role of an impedance. Note however that $H(\omega)$ is defined rather more generally than $\mathbf{Z}(\omega)$, having meaning for negative ω and for zero ω. The delta function in eqn (9.3) is necessary to ensure the correct initial condition with $h(t) = 0$ for $t < 0$ and if this delta function is removed it has the effect of adding a negative voltage to the output for all values of t.

As a second example let us consider an inductance L with input current $I(t)$ and output $V(t)$. This is less straightforward than the capacitor in that a current impulse produces a singularity in voltage that is not easily handled. Let us instead consider a step input so that

$$\left.\begin{array}{ll} I(t) = 0 & [t < 0] \\ = 1 & [t > 0] \end{array}\right\}. \tag{9.5}$$

The equation governing the behaviour of an inductance, and defining L, is

$$V(t) = L\frac{\mathrm{d}I}{\mathrm{d}t}$$

so that for the step input of eqn (9.5) we have a voltage response $b(t)$ given by:

$$b(t) = L\delta(t). \tag{9.6}$$

On Fourier transforming this we obtain

$$B(\omega) = L. \tag{9.7}$$

We may now obtain the transfer function $H(\omega)$ from eqn (7.20) giving

$$H(\omega) = i\omega L. \tag{9.8}$$

Once again we note that for $\omega > 0$ the transfer function is the same as the well known impedance

$$Z(\omega) = i\omega L. \tag{9.9}$$

These concepts can of course be extended to any L, R, C circuit and once the impedance has been calculated in the usual way by combining series and parallel elements we may derive $H(\omega)$ from $Z(\omega)$ since

$$H(\omega) = Z(\omega) \qquad [\omega > 0] \tag{9.10}$$

and

$$H(\omega) = H^*(-\omega) \qquad [\text{since } h(t) \text{ is real}]. \tag{9.11}$$

This only leaves a possible delta function at $\omega = 0$ to be determined, and this can be derived from an inspection of steady state requirements.

Energy and Power dissipation may be determined using the ideas derived in Sections 7.4 and 7.5. For instance eqn (7.32) shows that the energy dissipated depends on the energy spectrum of the applied signal and on the real part of the transfer function. This corresponds to the well-known fact that it is the resistive elements that cause energy dissipation.

9.3   Ladder Network Filters

A ladder network filter is constructed as in the diagram below. The impedances $\tfrac{1}{2}Z_1$, appearing at the ends of the network allow one to consider the circuit as made up from identical 'T' units with an impedance Z_2 on the stem and impedances $\tfrac{1}{2}Z_1$, on each side arm. We consider an alternating voltage V_i to be applied through an impedance Z_i to the input of the filter, and an impedance Z_0 to be connected across the output, the currents flowing at input and output being I_i and I_0 respectively.

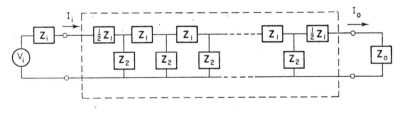

Fig. 9.1. A ladder network.

If we write

$$V_i = v_i \cos \omega t \qquad (9.12)$$

$$I_i = i_i \cos (\omega t + \phi_i) \qquad (9.13)$$

$$I_o = i_o \cos (\omega t + \phi_o) \qquad (9.14)$$

and introduce complex quntities $\mathbf{V}_i$, $\mathbf{I}_i$ and $\mathbf{I}_o$ as follows

$$\mathbf{V}_i = v_i \, e^{i\omega t} \qquad (9.15)$$

$$\mathbf{I}_i = i_i \, e^{i(\omega t + \phi_i)} \qquad (9.16)$$

$$\mathbf{I}_o = i_o \, e^{i(\omega t + \phi_o)} \qquad (9.17)$$

then the input and output currents may be derived from the following expressions (see for instance reference 5)

$$\mathbf{I}_o = \frac{\mathbf{V}_i}{Z_k + Z_i} \frac{e^{\Gamma n}(1 + T)}{1 - e^{2\Gamma n} ST} \qquad (9.18)$$

$$= \frac{\mathbf{V}_i}{Z_k + Z_i} \{ e^{\Gamma n} + T e^{\Gamma n} + ST e^{3\Gamma n} + ST^2 e^{3\Gamma n} + \cdots \} \qquad (9.19)$$

$$\mathbf{I}_i = \frac{\mathbf{V}_i}{Z_k + Z_i} \frac{1 + T e^{2\Gamma n}}{1 - e^{2\Gamma n} ST} \qquad (9.20)$$

$$= \frac{\mathbf{V}_i}{Z_k + Z_i} \{ 1 + T e^{2\Gamma n} + ST e^{2\Gamma n} + ST^2 e^{4\Gamma n} + S^2 T^2 e^{4\Gamma n} + \cdots \} \qquad (9.21)$$

In eqns (9.18)–(9.21) n is the number of repetitions of the basic 'T' unit in the network, and Γ, Z_k, S and T are related to the impedances present as below. We introduce the real quantities α and β for subsequent convenience.

$$\cosh \Gamma = \frac{Z_1 + 2Z_2}{2Z_2} = \cosh \alpha \cos \beta + i \sinh \alpha \sin \beta \qquad (9.22)$$

$$\Gamma = -\alpha - i\beta \qquad (9.23)$$

$$Z_k = \sqrt{Z_1 Z_2 + \tfrac{1}{4}Z_1{}^2} \qquad (9.24)$$

$$T = \frac{Z_k - Z_o}{Z_k + Z_o} \qquad (9.25)$$

$$S = \frac{Z_k - Z_i}{Z_k + Z_i}. \qquad (9.26)$$

Let us pause to interpret these results. The series expansions in eqns (9.19) and (9.21) make it appear as if the applied voltage produces a complex current $V_i/(Z_k + Z_i)$ at the input which is transmitted down the network being modified by a factor e^{Γ} at each 'rung' until at the far end it is reflected back, after modification by a reflection factor T, and transmitted back to the input. Here it is again reflected, after modification by a factor S, and so on. The overall current at input or output appears as the sum of the various multiply reflected components. From (9.25) and (9.26) we see that reflections are absent at the input or output when $Z_k = Z_i$ or $Z_k = Z_o$ respectively; under these conditions the input or output is said to be matched. Since Z_k is frequency dependent a system matched at one frequency will not necessarily be matched at another frequency. If reflections are absent, as with a matched output, then eqn (9.21) shows that $V_i = I_i(Z_k + Z_i)$, and we see that the filter has an input impedance of Z_k. This is the most direct interpretation for Z_k, and leads to its being known as the characteristic impedance of the filter. The constants α and β govern the change in amplitude and phase respectively of a component of current as it is transmitted from one rung to the next.

Having summarised the main properties of the filter in terms of standard alternating currents and voltages let us now treat it as a linear system using the concepts described in Chapter 7. Treating I_i as the input current, and I_o as the output, then by an argument closely analogous to that accompanying eqns (7.6)–(7.9) we find that the transfer function $H(\omega)$ is given by I_o/I_i so that from eqns (9.18) and (9.19) we get:

$$H(\omega) = \frac{I_o}{I_i} = \frac{e^{\Gamma n}(1 + T)}{1 + T e^{2\Gamma n}} \qquad [\omega > 0]. \qquad (9.27)$$

This represents in general a very complicated function of ω, so that a Fourier transform giving the impulse response $h(t)$ would be very difficult analytically. Let us however consider certain special cases. If the output is matched then $T = 0$ and we have, at the matching frequency,

$$H(\omega) = e^{\Gamma n}. \qquad (9.28)$$

If there are no resistive elements in the network so that Z_1 and Z_2 are purely imaginary then eqns (9.22) and (9.23) require that $\cosh \Gamma$ is real so that either $\sinh \alpha = 0$ or $\sin \beta = 0$. These two possibilities have a direct physical interpretation because they correspond to the current transmitted from rung to rung being unattenuated or attenuated respectively. In a long network the output is thus either unattenuated or very nearly zero. Regions of frequency for which the former holds are called pass-bands, and those for which the latter holds are called stop-bands. The following results are easily derived from eqns (9.22), (9.23) and (9.28).

(a) Pass-band

$$\left| \frac{\mathbf{Z}_1 + 2\mathbf{Z}_2}{2\mathbf{Z}_2} \right| < 1, \quad \alpha = 0, \quad \cos \beta = \frac{\mathbf{Z}_1 + 2\mathbf{Z}_2}{2\mathbf{Z}_2} \tag{9.29}$$

$$H(\omega) = e^{-in\beta} \quad \text{(for matched output)} \tag{9.30}$$

$$\mathbf{Z}_k \quad \text{is real} \tag{9.31}$$

(b) Stop-band

$$\left| \frac{\mathbf{Z}_1 + 2\mathbf{Z}_2}{2\mathbf{Z}_2} \right| > 1 \tag{9.32}$$

$$\cosh \alpha = \frac{\mathbf{Z}_1 + 2\mathbf{Z}_2}{2\mathbf{Z}_2} \cos \beta \tag{9.33}$$

$$\beta = 0, \ \pm\pi, \ \pm2\pi, \ \pm3\pi, \ \ldots \tag{9.34}$$

$$H(\omega) = e^{-n\alpha} \cos \beta \quad \text{(for matched output)} \tag{9.35}$$

$$\mathbf{Z}_k \quad \text{is pure imaginary.} \tag{9.36}$$

Continuing with yet more simplifications, the following special forms for $\mathbf{Z}_1$ and $\mathbf{Z}_2$ give useful forms of frequency response. We show the characteristics diagramatically by plotting α and β against frequency. The names low-pass, high-pass, band-pass and band-stop are clearly related to that region of the frequency axis corresponding to a pass-band.

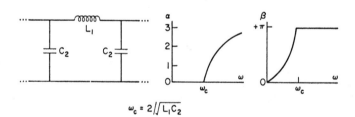

$$\omega_c = 2/\sqrt{L_1 C_2}$$

FIG. 9.2. Low-pass filter.

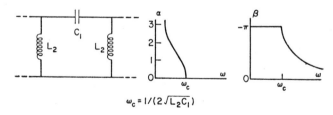

$$\omega_c = 1/(2\sqrt{L_2 C_1})$$

FIG. 9.3. High-pass filter.

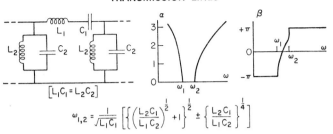

$$\omega_{1,2} = \frac{1}{\sqrt{L_1 C_1}} \left[\left\{ \left(\frac{L_2 C_1}{L_1 C_2} \right)^{\frac{1}{2}} + 1 \right\}^{\frac{1}{2}} \pm \left\{ \frac{L_2 C_1}{L_1 C_2} \right\}^{\frac{1}{4}} \right]$$

FIG. 9.4. Band-pass filter.

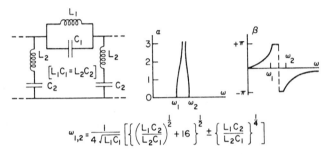

$$\omega_{1,2} = \frac{1}{4\sqrt{L_1 C_1}} \left[\left\{ \left(\frac{L_1 C_2}{L_2 C_1} \right)^{\frac{1}{2}} + 16 \right\}^{\frac{1}{2}} \pm \left\{ \frac{L_1 C_2}{L_2 C_1} \right\}^{\frac{1}{4}} \right]$$

FIG. 9.5. Band-stop filter.

It might be thought that if a signal has a transform $F(\omega)$ which is zero outside the pass band of a filter then the signal will pass through unaltered. Such is not the case however since the different frequency components can suffer different phase changes so that the output is distorted, though it is true that the power or energy spectrum of the signal remains unaltered. We return to this discussion in Sections 9.5–9.7 where we will show that for distortionless transmission of signals we require that $\beta \propto \omega$, a condition which is approximately satisfied for low frequencies in a low pass filter but which cannot be satisfied exactly for all frequencies within a pass band.

9.4   Transmission Lines

We will consider a transmission line consisting of two parallel conductors between which there is a conductance G and capacitance C per unit length, and along which there is an inductance L and resistance R per unit length (including both conductors). We consider an alternating voltage V_i to be applied through an impedance Z_i to the input of the line, and an impedance

Z_o to be connected across the output, the currents flowing at input and output being I_i and I_o as in the diagram.

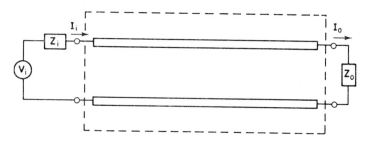

FIG. 9.6. Transmission line.

The quantities V_i, I_i and I_o are related by a set of equations closely analogous to those governing the behaviour of a ladder filter and indeed eqns (9.12)–(9.21) may be applied directly to the transmission line provided we reinterpret the quantities Γ, Z_k, S, T and n as follows. n now represents the length of the line (and is thus a continuous variable instead of an integer), and Γ, α and β are quantities that refer to unit length of line. We have to replace eqns (9.22)–(9.26) by:

$$\Gamma = -\alpha - i\beta = -\sqrt{(R + i\omega L)(G + i\omega C)} \tag{9.37}$$

$$Z_k = \left(\frac{R + i\omega L}{G + i\omega C}\right)^{\frac{1}{2}} \tag{9.38}$$

$$S = \frac{Z_k - Z_i}{Z_k + Z_i} \tag{9.39}$$

$$T = \frac{Z_k - Z_o}{Z_k + Z_o}. \tag{9.40}$$

The remarks made in Section 9.3 about matching, reflections and the interpretation of characteristic impedance apply here also.

The transfer function comes to be

$$H(\omega) = \frac{I_o}{I_o} = \frac{e^{\Gamma l}(1 + T)}{1 + T\,e^{2\Gamma l}} \qquad [\omega > 0] \tag{9.41}$$

which for a matched output reduces to

$$H(\omega) = e^{\Gamma l} \quad [\omega = \text{matching frequency}]. \tag{9.42}$$

We have here represented the length of the line by l, as being a more appropriate symbol than n.

As an approximation often met in practice at high frequencies let us consider the case

$$G = 0 \tag{9.43}$$

$$R \ll \omega L. \tag{9.44}$$

It then follows that $\alpha \approx (R/2)\sqrt{C/L}$ and $\beta \approx \omega\sqrt{LC}$ so that with a matched output

$$H(\omega) = \exp\left[-(lR/2)\sqrt{C/L}\right] \exp\left(-il\omega\sqrt{LC}\right). \tag{9.45}$$

This form for the transfer function, with phase lag proportional to frequency, is just that required to ensure that signals are transferred without distortion (see Section 9.5) corresponding to the fact that waves of all frequencies travel down the line at the same rate. On Fourier transforming $H(\omega)$ we obtain the impulse response

$$h(t) = \exp\left[-(lR/2)\sqrt{C/L}\right] \delta(t - l\sqrt{LC}) \tag{9.46}$$

showing that signals travel along the line at a speed $1/\sqrt{LC}$. For two parallel wires in free space this velocity equals the velocity of light. If however the line is 'loaded' to increase its inductance above this value (by shielding the wires with high permeability material or by inserting inductances at intervals) then the velocity is reduced (and also the attenuation is reduced). The distortionless property is retained, except that the method of adding inductances at intervals does introduce a high frequency cut-off, as the system becomes effectively a special type of ladder network filter.

9.5   Ideal Distortionless Filter

Having considered actual systems in Sections 9.1–9.4 let us for the remainder of this chapter consider some formal properties of filters which are independent of their internal construction.

An ideal distortionless filter is one which will pass any signal unchanged in shape and amplitude, but possibly delayed. Thus if $f(t)$ and $g(t)$ represent input and output we have:

$$g(t) = f(t - t_0) \tag{9.47}$$

where t_0 is the delay. On Fourier transforming we obtain

$$G(\omega) = F(\omega)\, e^{-i\omega t_0} \tag{9.48}$$

so that the transfer function must be

$$H(\omega) = e^{-i\omega t_0}. \tag{9.49}$$

Such a filter is called a linear phase shift filter because it introduces a phase shift proportional to frequency. The result is physically plausible because if harmonic signals of all frequencies are to be delayed by the same amount then clearly those of high frequency must be delayed by a larger fraction of one period of oscillation than those of low frequency, so that the phase delay is proportional to frequency.

9.6   Low-Pass Filters

An ideal low-pass filter is specified by the transfer function:

$$\left. \begin{aligned} H(\omega) &= e^{-i\omega t_0} \quad [|\omega| < \omega_c] \\ &= 0 \quad\quad [|\omega| \geqslant \omega_c] \end{aligned} \right\}. \tag{9.50}$$

It is easily verified that a signal whose transform $F(\omega)$ is zero for $|\omega| > \omega_c$ is passed through undistorted, whereas a signal with $F(\omega)$ zero for $|\omega| < \omega_c$ is not passed through at all.

The impulse response is obtained on Fourier inverting $H(\omega)$ to give

$$h(t) = \frac{\sin \omega_c (t - t_0)}{\pi(t - t_0)} \tag{9.51}$$

and we notice that such a filter is not possible in practice since $h(t)$ is not zero for $t < 0$ so that $h(t)$ is not causal. However if sufficient delay exists ($t_0 \gg 1/\omega_c$) then one may approach very close to the ideal since $h(t)$ in expression (9.51) is now very small for $t < 0$.

The causality requirement, namely $h(t) = 0$ for $t < 0$, is clearly a law of nature since effect must follow cause and it is of some interest to know what is the corresponding restriction of $H(\omega)$. A determination of the necessary and sufficient conditions that $H(\omega)$ must satisfy in order that $h(t)$ shall be causal is not easy and is discussed for instance in reference 12. It is found that the real and imaginary parts of $H(\omega)$ are related in a special way, and that if $h(t)$ is square integrable then a necessary (but not sufficient) condition for $h(t)$ to be causal is that

$$\int_{-\infty}^{+\infty} \frac{|\ln|H(\omega)||}{1 + \omega^2} \, d\omega < \infty.$$

This condition is known as the Paley–Wiener condition, and from it we may see for instance that an ideal band-stop filter is not physically realizable. Some consequences of causality are discussed in Appendix U.

A signal whose transform $F(\omega)$ exists both with $\omega_0 > \omega_c$ and $\omega < \omega_c$ is distorted by a low pass filter due to 'having its high frequency components removed'. If $f(t)$ is periodic with period $T = 2\pi/\omega$ then $F(\omega)_0$ consists of delta functions at multiples of the fundamental frequency ω_0 so that:

$$F(\omega) = \sum_{n=-\infty}^{+\infty} a_n \delta(\omega - n\omega_0) \qquad [a_n = a_{-n}^*]. \tag{9.52}$$

As ω_c is reduced from a high value so the harmonics are cut out one by one. Indeed for $|\omega_0| < |\omega_c| < 2|\omega_0|$ the only components allowed through are the fundamental and a D.C. component, and it is easily verified that in this case the output is

$$g(t) = (a_1/\pi) \cos \omega(t - t_0) + (a_0/2\pi). \tag{9.53}$$

As an extreme example of this, a regular sequence of impulses may thus be 'converted' to an harmonic wave (plus D.C. bias) by a suitable low-pass filter.

The 'actual' low-pass filter discussed in Section 9.3 is not ideal, since the phase shift is not linear. From eqn (9.29) we see that in this case

$$\cos \beta = 1 + \frac{Z_1}{2Z_2} = 1 - \tfrac{1}{2}\omega^2 L_1 C_2. \tag{9.54}$$

However at low frequencies with $\omega \ll 1/\sqrt{L_1 C_2}$ it follows that

$$\beta \approx \omega\sqrt{L_1 C_2} \tag{9.55}$$

so that the system is linear in phase and under matched conditions

$$H(\omega) \approx \exp\left(-in\omega\sqrt{L_1 C_2}\right) \qquad [\omega^2 LC \ll 1]. \tag{9.56}$$

We notice that the delay is $t_0 = n\sqrt{L_1 C_2}$, and notice further that the reflection interpretation of eqns (9.19) and (9.21) is confirmed since the terms $e^{\Gamma n}$, $e^{3\Gamma n}$, etc correspond to echoes with increasing time delays.

The phenomenon of echoes is apparent in any small deviation from ideality if the deviation is well represented by a Fourier series expansion. For instance consider the filter characterised by amplitude distortion with $H(\omega)$ given by:

$$\begin{aligned} H(\omega) &= A(\omega)\,e^{-i\omega t_0} & [\omega \leqslant \omega_c] \\ &= 0 & [\omega > \omega_c] \end{aligned} \Bigg\} \tag{9.57}$$

where $A(\omega)$ instead of being independent of frequency is a function of ω. Since $A(\omega)$ is an even function existing in the range $-\omega_c < \omega < \omega_c$ we may write $A(\omega)$ as a cosine Fourier series (see eqns (1.13)–(1.16)) of the form

$$A(\omega) = a_0 + \sum_{n=1}^{\infty} a_n \cos\left(\frac{n\pi\omega}{\omega_c}\right) \qquad (9.58)$$

where

$$a_n = \frac{2}{\omega_c}\int_0^{\omega_c} A(\omega) \cos\left(\frac{n\pi\omega}{\omega_c}\right) d\omega$$

$$a_0 = \frac{1}{\omega_c}\int_0^{\omega_c} A(\omega)\, d\omega. \qquad (9.59)$$

We leave it as an exercise for the reader to show now that if the impulse response from the related ideal filter is written $h_0(t)$, then the impulse response to the above filter may be written as the convolution

$$h(t) = h_0(t) \otimes \left[a_0\delta(t) + \sum_{n=1}^{\infty} \frac{a_n}{2}\left\{ \delta\left(t - \frac{n\pi}{\omega_c}\right) + \delta\left(t + \frac{n\pi}{\omega_c}\right)\right\}\right]. \qquad (9.60)$$

This shows that the ideal response is accompanied by delayed and anticipatory echoes. The anticipatory echoes are not all necessarily non-physical if $h_0(t)$ contains delay.

We may alternatively consider deviations from ideality of the following sort (phase distortion). Consider the filter characterised by

$$\left.\begin{array}{ll} H(\omega) = e^{i\phi(\omega)} & [|\omega| \leqslant \omega_c] \\ = 0 & [|\omega| > \omega_c] \end{array}\right\} \qquad (9.61)$$

where $\phi(\omega)$ instead of being linear in ω may be written as a linear term plus a Fourier series expansion of the deviation from linearity

$$\phi(\omega) = -\omega t_0 - \left\{\sum_{n=1}^{\infty} \phi_n \sin\left(\frac{n\pi\omega}{\omega_c}\right)\right\} \qquad [|\omega| < \omega_c] \qquad (9.62)$$

where

$$\phi_n = -\frac{2}{\omega_c}\int_0^{\omega_c}(\phi(\omega) + \omega t_0) \sin\left(\frac{n\pi\omega}{\omega_c}\right) d\omega. \qquad (9.63)$$

If the coefficients ϕ_n are small compared with unity then we again leave it as an exercise for the reader to verify that if $h_0(t)$ is the impulse response for the associated ideal filter, then for the actual filter we have the convolution

$$h(t) = h_0(t) \otimes \left[\delta(t) + \sum_{n=1}^{\infty}\left\{ \frac{\phi_n}{2}\delta\left(t - \frac{n\pi}{\omega_c}\right) - \frac{\phi_n}{2}\delta\left(t + \frac{n\pi}{\omega_c}\right)\right\}\right]. \qquad (9.64)$$

This shows once again the phenomenon of delayed and anticipatory echoes.

9.7 Band-Pass Filters

An ideal band-pass filter is one with the following transfer function

$$H(\omega) = e^{-i\omega t_0} \quad [\omega_1 < |\omega| < \omega_2] \\ = 0 \qquad \text{otherwise} \qquad \Bigg\} \qquad (9.65)$$

where the frequency range between ω_1 and ω_2 is the band-pass region (which does not include the origin).

It is easily verified that if the input signal $f(t)$ has a transform $F(\omega)$ which is zero outside the band-pass then the signal passes through undistorted, and delayed by t_0.

The impulse response is obtained by Fourier inversion of $H(\omega)$ and is

$$h(t) = \frac{2}{\pi} \frac{\cos \omega_0(t - t_0) \sin \omega_c(t - t_0)}{(t - t_0)}. \qquad (9.66)$$

We have here used ω_0 to represent the centre frequency of the pass-band, and $2\omega_c$ to represent the width of the pass-band. We notice that $h(t)$ is non-causal so that the ideal pass-band filter is not possible in practice, though if the delay is large $(t_0 \gg 1/\omega_c)$ then $h(t)$ will be small for $t < 0$.

If the band-pass is narrow so that $\omega_c \ll \omega_0$ then eqn (9.66) corresponds to a carrier wave at frequency ω_0, amplitude modulated according to the term $\{\sin \omega_c(t - t_0)/(t - t_0)\}$. Indeed it is plausible to expect that the output will approximate to a modulated carrier wave at frequency ω_0 whatever the input, if the band-pass is narrow. This receives rigorous justification in the following discussion.

We show now that the impulse response of a band-pass filter can always be written in the form

$$h(t) = 2h_p(t) \cos \omega_0 t + 2h_q(t) \sin \omega_0 t. \qquad (9.67)$$

This is not limited to narrow band filters, nor to ideal band-pass filters, and ω_0 now refers to any frequency chosen for reference within the pass band. In eqn (9.67) $h_p(t)$ and $h_q(t)$ are the impulse responses of two low-pass filters related to the band-pass filter. These low-pass filters are known as the in-phase and quadrature filters since they determine the cosine and sine terms respectively in eqn (9.67). The transfer functions $H_p(\omega)$ and $H_q(\omega)$ of the in-phase and quadrature filters are so chosen that the transfer function $H(\omega)$ of the band-pass filter may be built up from $H_p(\omega)$ and $H_q(\omega)$ by 'shifting them away from the origin' and then adding and subtracting them appropriately as follows:

$$H(\omega) = H_p(\omega - \omega_0) - iH_q(\omega - \omega_0) + H_p(\omega + \omega_0) + iH_q(\omega + \omega_0). \qquad (9.68)$$

Some thought will show that $H_p(\omega)$ and $H_q(\omega)$ may be constructed from $H(\omega)$ as follows:

$$H_p(\omega) = \tfrac{1}{2}\{H(\omega + \omega_0) + H(\omega - \omega_0)\} \qquad [-\omega_0 < \omega < +\omega_0] \qquad (9.69)$$

$$H_q(\omega) = (i/2)\{H(\omega + \omega_0) - H(\omega - \omega_0)\} \qquad [-\omega_0 < \omega < +\omega_0] \qquad (9.70)$$

$$H_p(\omega) = \tfrac{1}{2}H(\omega + \omega_0) \qquad\qquad\qquad\qquad [\omega > \omega_0] \qquad (9.71)$$

$$H_q(\omega) = (i/2)H(\omega + \omega_0) \qquad\qquad\qquad\quad [\omega > \omega_0] \qquad (9.72)$$

$$H_p(\omega) = \tfrac{1}{2}H(\omega - \omega_0) \qquad\qquad\qquad\qquad [\omega < -\omega_0] \qquad (9.73)$$

$$H_q(\omega) = (-i/2)H(\omega - \omega_0) \qquad\qquad\qquad [\omega < -\omega_0]. \qquad (9.74)$$

When defined as above all the transfer functions have the correct symmetry to ensure that the impulse responses are real functions. Fourier inversion of (9.68) leads straightway to the desired result, eqn (9.67).

The effect that a band-pass filter has on an amplitude modulated signal can conveniently be discussed terms of the in-phase and quadrature low-pass filters. If the input can be written

$$f(t) = m(t) \cos \omega_0 t \qquad (9.75)$$

then the output comes to be (under conditions discussed below, eqn (9.77))

$$g(t) = n_p(t) \cos \omega_0 t + n_q(t) \sin \omega_0 t \qquad (9.76)$$

where $n_p(t)$ and $n_q(t)$ are the outputs from the in-phase and quadrature filters for an input $m(t)$. Equation (9.76) may be readily derived by first expressing $G(\omega)$ as the product of $F(\omega)$ and $H(\omega)$, deriving $F(\omega)$ from eqn (9.75) as $F(\omega) = \tfrac{1}{2}\{M(\omega - \omega_0) + M(\omega + \omega_0)\}$ and using $H(\omega)$ as in eqn (9.68). The resulting expression for $G(\omega)$ reduces to give eqn (9.76) provided that certain cross terms such as $M(\omega - \omega_0)H_p(\omega + \omega_0)$ are zero. This will be so provided

$$2\omega_0 > \omega_m + \omega_{pq} \qquad (9.77)$$

where ω_{pq} is the cut-off frequency of the in-phase and quadrature filters, and ω_m is an effective limiting frequency such that $M(\omega) = 0$ for $|\omega| > \omega_m$. We have tacitly assumed the carrier frequency to lie in the pass-band, but in fact eqns (9.67)–(9.77) still apply if ω_0 is outside the pass-band the only change being that the in-phase and quadrature filters cease to be low-pass filters. Equation (9.76) can thus be used to discuss a situation in which the carrier frequency perhaps lies outside the pass-band, but $F(\omega)$ and $H(\omega)$ still overlap due to the broadening of $F(\omega)$ caused by the amplitude modulation. Equation (9.76) can be interpreted by imagining that the carrier wave passes through unaltered (except for the possible creation of a component $\pi/2$ out of phase, represented by the sine term) but that the modulating function is subject to filtering.

Chapter 10

The Retrieval of Information from Noise

10.1 Introduction

We will start with a survey of some typical problems in information retrieval. A familiar problem is the separation of music or speech from hiss or background noise when listening to a gramophone record or a distant radio station. This is essentially the separation of one random signal from another one and a well-known solution is to introduce a filter which lets through the frequency band covered by the wanted signal but cuts out the frequency components in the noise. If the power spectra overlap then a complete separation is not possible, and we have to consider what is the best compromise. We return to such a discussion in Section 10.6 with a consideration of the so called Wiener–Hopf condition.

Another common problem is the measurement of a periodic signal buried in noise as for instance arises in analysing the complicated sound produced by various pieces of machinery working together. If the frequencies and amplitudes of periodic components in the sound can be measured then this is a first step towards tracing their causes. An easier problem is the retrieval of a periodic signal from random noise when a reference waveform is available with which the signal is known to be synchronous. This arises in A.C. testing when an alternating voltage is applied to a system and a weak voltage induced elsewhere is to be measured in the presence of noise. Another context is that of impulse testing when a repeated delta function impulse is applied to a system, and the repeated response is to be measured. The problem is further simplified if the shape of the signal to be retrieved is known, and one merely wants to determine the phase lag relative to a reference signal. Such is the case in pulsed radar when the amplitude and delay of an echo or reflected signal are to be determined in the presence of noise.

Sometimes the shape of the signal to be extracted is known (at least approximately) but no synchronising signal exists. This is the case with the seismic measurements related to nuclear explosions and earthquakes. The signals associated with each have their known characteristics and the problem is to determine in the presence of noise whether such a signal has occurred and which type it is. The problem of deciding which of several known types of signal has arrived occurs in digital communication systems, the problem being often simplified in this case by the arrival time being known. Morse code provides a simple binary example. During each of successive equal periods there is either a uniform signal or no signal and a device is required which will distinguish between the alternatives in the presence of noise. We discuss problems such as this in Section 10.7 with a consideration of the matched filter.

The various situations are sometimes classified according to whether the signal is *coherent* or *incoherent* (i.e. as to whether a synchronising waveform is available or not) and according to whether the signal is structurally *determinate* or *indeterminate* (i.e. as to whether the shape of the signal is known in detail, or whether only statistical information such as the power spectrum is known).

In Sections 10.2–10.5 we describe various instruments which are commonly used in processing signals, and in Sections 10.6 and 10.7 we discuss the more mathematical aspects of optimum performance.

10.2   The Integrator

An integrating or smoothing device simply smooths out rapid fluctuations in a waveform, this process being equivalent to taking a running average over a certain period of time. The process is alternatively equivalent to passing the signal through a form of low-pass filter which cuts out high frequency components relative to low frequency ones. Regarded as a device for the extraction of signal from noise it performs the humble but useful function of retrieving a constant D.C. signal from noise since ideally the noise components are averaged to zero.

If the input to an integrator is $f(t)$ and the output $g(t)$, then one form of integrating action is represented by

$$g(t) = \frac{1}{T} \int_{t-T}^{t} f(t_1)\, dt_1 \qquad (10.1)$$

so that the output at time t is equal to the average of the input during the period T immediately preceeding t. We can write eqn (10.1) alternatively as a convolution

$$g(t) = \int_{-\infty}^{+\infty} f(t_1)\, h(t - t_1)\, dt_1 \tag{10.2}$$

$$= f(t) \otimes h(t)$$

where $h(t)$ is a gating function such that

$$\left. \begin{array}{ll} h(t) = \dfrac{1}{T} & [0 \leqslant t \leqslant T] \\[2mm] = 0 & \text{otherwise.} \end{array} \right\} \tag{10.3}$$

Comparison of eqns (10.2) and (7.2) shows that we may regard the integrator as a linear system with impulse response $h(t)$. In the frequency domain we can regard this system as a filter with a transfer function $H(\omega)$ given by the Fourier transform of $h(t)$ so that from eqn (2.49)

$$H(\omega) = \frac{\sin(\omega T/2)}{(\omega T/2)}\, e^{-i\omega T/2}. \tag{10.4}$$

A simple RC circuit is often used as an integrator, the integration being exponential, by which we mean that $h(t)$ is not as in eqn (10.3), but is of the form

$$\left. \begin{array}{ll} h(t) = \dfrac{1}{T} e^{-t/T} & [\text{for } t \geqslant 0] \\[2mm] = 0 & [t < 0] \end{array} \right\} \tag{10.5}$$

so that from eqn (2.40)

$$H(\omega) = \frac{1}{T} \left\{ \frac{(1/T) - i\omega}{(1/T)^2 + \omega^2} \right\}. \tag{10.6}$$

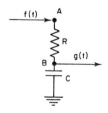

FIG. 10.1. An integrating circuit.

We call T the integration time constant, and it is equal to the product RC in the simple circuit discussed here. For simplicity we have normalised $h(t)$ so as to have an integrated value of unity, and we shall assume $h(t)$ to be so normalised during the remainder of this chapter.

The integrating circuit is shown in Fig. 10.1; a voltage $f(t)$ is applied at A and an output voltage $g(t)$ is measured at B. The action of such a circuit may be described in several ways. For RC very large the voltage at B is proportional to the charge on the condenser which is proportional to the integrated value of the current which has flowed into it which in turn is proportional to the integrated value of the voltage applied at A. Hence its name as an integrator. For RC not necessarily large we may proceed by considering the response to a unit step voltage applied at A. Elementary electrical theory shows that the voltage at B rises exponentially to the value unity, with a step response $b(t)$ given by

$$b(t) = 1 - e^{-t/RC} \qquad (10.7)$$

so that on differentiating (see eqn (7.17)) we get the impulse response

$$h(t) = \frac{1}{RC} e^{-t/RC} \qquad [t > 0] \qquad (10.8)$$

in agreement with eqn (10.5). Alternatively we may note that $H(\omega)$, at least for $\omega > 0$, may be obtained from the ratio of input to output for an alternating voltage, as in eqns (7.6)–(7.8). Thus, using the complex impedances $- i/\omega C$ and R for the capacitor and resistor respectively, we obtain

$$H(\omega) = \frac{- i/\omega C}{R - (i/\omega C)} \qquad (10.9)$$

which reduces to give eqn (10.6) as it should do.

As a final consideration let us work out the reduction in mean square amplitude of noise fluctuations when they pass through an integrator. In the special case of white noise having a power spectrum of constant magnitude up to some cut-off frequency ω_c, we may guess that the noise will be reduced by an amount proportional to the ratio of ω_c to the equivalent filter bandwidth. The power spectra of the input and output are related by (see eqn (7.38))

$$P_g(\omega) = P_f(\omega) |H(\omega)|^2 \qquad (10.10)$$

so that we may use Parseval's theorem, eqn (4.10), to give the mean square value of the output as

$$\langle g^2(t) \rangle = \frac{1}{2\pi} \int_{-\infty}^{+\infty} P_f(\omega)\,|H(\omega)|^2\,d\omega \qquad (10.11)$$

while the mean square value at the input (see eqn (4.10)) is

$$\langle f^2(t) \rangle = \frac{1}{2\pi} \int_{-\infty}^{+\infty} P_f(\omega)\,d\omega. \qquad (10.12)$$

If the input is white noise so that $P_f(\omega)$ is constant up to some frequency ω_c which is large compared with the frequencies involved in $H(\omega)$ then we obtain from eqns (10.11) and (10.12)

$$\frac{\langle g^2(t) \rangle}{\langle f^2(t) \rangle} = \frac{\int_{-\infty}^{+\infty} |H(\omega)|^2\,d\omega}{2\omega_c} \qquad (10.13)$$

$$= \frac{\pi}{\omega_c} \int_{-\infty}^{+\infty} h^2(t)\,dt\,. \qquad (10.14)$$

For the RC circuit with $h(t)$ as in eqn (10.8) this reduces to a ratio $(\pi/2\omega_c RC)$, which is of the expected form.

10.3 The Phase Sensitive Detector

This device is useful for measuring the magnitude of a harmonic signal buried in noise when a synchronous wave form is available. The input (signal plus noise) is multiplied by a harmonic reference waveform and the product is integrated or smoothed. Since the signal to be extracted and the reference waveform are in phase their product is always positive and the output after smoothing is a steady positive output. The product of the noise signal with the reference waveform however averages to zero for sufficiently long averaging times, so that ideally one obtains a steady output proportional to the magnitude of the harmonic signal. In practice the output will be fluctuating about this steady value since with a finite integrating time the noise does not average exactly to zero. In case the signal is shifted in phase relative to the reference waveform it can be arranged that the phase of the reference is adjustable and one then searches for maximum response as one varies the phase of the reference waveform.

For an input $f(t)$ consisting of a harmonic signal at frequency ω_0 plus noise $n(t)$,

$$f(t) = A \cos \omega_0 t + n(t) \qquad (10.15)$$

the output may be written, using eqn (10.2)

$$g(t) = \int_{-\infty}^{+\infty} \{f(t_1) \cos \omega_0 t_1\} h(t - t_1) \, dt_1 \tag{10.16}$$

where $h(t)$ is the characteristic of the integrating circuit, typically either a rectangular gate or an exponential as in eqns (10.3) or (10.5). For very long integrating time constants we see from eqn (10.16) that $g(t)$ represents the cross-correlation, for zero delay, between the input $f(t)$ and the reference signal $\cos \omega_0 t$. For finite integrating times the output becomes the cross-correlation between a truncated version of $f(t)$ with the reference waveform. An alternative way of regarding the process is to consider it simply as an integrator into which is fed a signal $f(t) \cos \omega_0 t$. The cosine term has the effect of shifting all the frequency components in $f(t)$ by an amount $\pm \omega_0$, so that a D.C. component is produced from any component in $f(t)$ at $\omega = \omega_0$. This D.C. component is then allowed through the integrator (regarded as a low-pass filter).

Regarding the device as an integrator with input $f(t) \cos \omega_0 t$ we may use eqn (10.11) to give an expression for the mean square fluctuation at the output. Since the power spectrum of $n(t) \cos \omega_0 t$ is the same as that of $n(t)$ (provided the power spectrum $P_n(\omega)$ stretches up to frequencies well in excess of ω_0, see eqn (6.33)) the mean square output $\langle n^2_0(t) \rangle$ when noise alone is fed into the input will be

$$\langle n_0^2(t) \rangle = \frac{1}{2\pi} \int_{-\infty}^{+\infty} P_n(\omega) \, |H(\omega)|^2 \, d\omega . \tag{10.17}$$

Thus if $n(t)$ is white noise up to some cut-off frequency ω_c we have from eqn (10.14)

$$\frac{\langle n^2_0(t) \rangle}{\langle n^2(t) \rangle} = \frac{\pi}{\omega_c} \int_{-\infty}^{+\infty} h^2(t) \, dt \tag{10.18}$$

$$= \frac{\pi}{2\omega_c RC} \qquad \text{[for an } RC \text{ integrator]} . \tag{10.19}$$

The steady output due to the signal will be $A/2$ since the mean value of $\cos^2 \omega_0 t$ is one half.

10.4 The Boxcar Detector

This device is useful for determining the shape of a repetitive waveform buried in noise when a synchronising pulse is available. In effect it stores several successive versions of the waveform and presents a function equal to the average waveform, random variations being averaged out. There is a certain similarity with the simple expedient of looking at a triggered waveform on an oscilloscope with a long persistence screen; here also fluctuations and jitter are smoothed out leaving one with the required average waveform.

The actual mode of operation is indicated in Fig. 10.2. The upper waveform represents the incoming signal (drawn without noise for convenience), and the lower waveform represents a repetitive gating signal

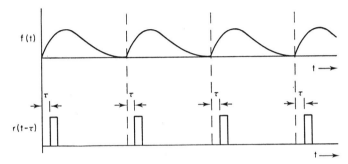

Fig. 10.2. The action of a boxcar detector.

of variable delay τ which 'selects' equivalent portions of successive versions of the input. The product of these two waveforms is averaged over several cycles using an integrating circuit characterised as usual by a function $h(t)$, so that one is provided with an averaged value of $f(t)$ at a delay τ relative to the synchronising pulse. By varying τ one is thus able to build up a picture of the signal.

The output may be written

$$g(t, \tau) = \int_{-\infty}^{+\infty} f(t_1) \, r(t_1 - \tau) \, h(t - t_1) \, dt_1. \qquad (10.20)$$

Regarded as a function of τ the output, at fixed t, can be regarded as the cross-correlation between a truncated version of $f(t_1)$ and the gating waveform. Indeed in those versions of the device which utilise digital techniques a bank of functions $r(t - \tau)$ with various values of τ are available simultaneously, so that at any time t a display is given of $g(t, \tau)$ as a function of τ. This display has the shape of the waveform to be extracted.

10.5 The Correlator

This device is useful amongst other things for detecting the presence of a waveform of known shape amongst noise, though its versatility is such that the phase sensitive detector and boxcar detector represent essentially special uses of a correlator.

Provision is made for two inputs to be applied, $a(t)$ and $b(t)$. The device delays one signal by an adjustable amount τ, then multiplies the signals together and finally integrates over some suitable period according to a function $h(t)$. The output $g(t)$ is thus

$$g(t, \tau) = \int_{-\infty}^{+\infty} a(t_1 - \tau) \, b(t_1) \, h(t - t_1) \, dt_1. \tag{10.21}$$

As is the case with boxcar detectors some instruments possess a single delay unit and $g(t, \tau)$ as a function of τ has to be built up from successive measurements with various values of τ, whilst the more expensive instruments have an array of delay units available for simultaneous use so that at any time t a display of $g(t, \tau)$ versus τ is possible.

From eqn (10.21) we see that since $a(t)$ and $b(t)$ are real functions the output is simply the cross-correlation between $a(t_1)$ and $b(t_1) h(t - t_1)$, that is between $a(t_1)$ and a truncated portion of $b(t_1)$ in the neighbourhood of $t_1 = t$. As the integrating time constant is lengthened so the output approaches the ideal cross-correlation $R_{ab}(\tau)$. Equivalently the same result is obtained for a finite integrating time by averaging $g(t, \tau)$ over all values of t. This may be shown as follows, using triangular brackets to represent an average over all values of t:

$$\langle g(t, \tau) \rangle = \lim_{T \to \infty} \frac{1}{2T} \int_{-T}^{+T} \left\{ \int_{-\infty}^{+\infty} a(t_1 - \tau) \, b(t_1) \, h(t - t_1) \, dt_1 \right\} dt \tag{10.22}$$

$$= \lim_{T \to \infty} \frac{1}{2T} \int_{-\infty}^{+\infty} \left\{ \int_{-T}^{+T} a(t_1 - \tau) \, b(t_1) \, h(t - t_1) \, dt_1 \right\} dt \tag{10.23}$$

$$= \lim_{T \to \infty} \frac{1}{2T} \int_{-T}^{+T} \left\{ \int_{-\infty}^{+\infty} a(t_1 - \tau) \, b(t_1) \, h(t - t_1) \, dt \right\} dt_1 \tag{10.24}$$

$$= \lim_{T \to \infty} \frac{1}{2T} \int_{-T}^{+T} a(t_1 - \tau) \, b(t_1) \, dt_1 \tag{10.25}$$

$$= R_{ab}(\tau). \tag{10.26}$$

If the same signal is fed into both inputs then the device produces the autocorrelation of the input. This is useful in searching for periodic signals buried in noise when no synchronous signal is available. The autocorrelation consists of the sum of the separate autocorrelations of the noise and of the periodic signal (see Section 6.4). Since the autocorrelation function of a periodic signal is periodic up to all values of τ, whilst for noise $n(t)$ the autocorrelation settles to a steady value equal to $\{\langle n(t) \rangle\}^2$ at large τ (see eqn (6.12)) we may discern the periodic signal by the ripple in $R_f(\tau)$. This is illustrated in Fig. 10.3 for a periodic signal buried in noise $n(t)$.

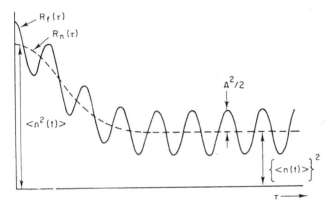

FIG. 10.3. The autocorrelation $R_f(\tau)$ for $f(t) = A \cos \omega t + n(t)$.

Another application lies in detecting the amplitude and time of arrival of a signal of known shape buried in noise. The input (signal plus noise) is fed in as input $b(t)$ while a waveform having the shape of the expected signal is fed in as input $a(t)$. The output $g(\tau)$ has a maximum value at a value τ equal to the delay between the reference and its 'echo'.

A correlator provides an ingenious way of measuring the impulse response of a system. White noise is fed into the system under test and the cross-correlation between this input and the output is measured. This cross-correlation is equal to the impulse response of the system. The proof of this is straightforward if we work in the frequency domain. Let $f(t)$ and $g(t)$ be finite power input and output signals relating to the system under test which has an impulse response $h(t)$. Let a subscript T refer to the corresponding quantities for a truncated portion $f_T(t)$ of the input. Then according to eqns (7.2) and (7.3)

$$g_T(t) = f_T(t) \otimes h(t)$$

$$G_T(\omega) = F_T(\omega) H(\omega)$$

so that
$$F_T^*(\omega)\, G_T(\omega) = F_T^*(\omega)\, F_T(\omega)\, H(\omega) \tag{10.27}$$

$$\lim_{T \to \infty} \frac{1}{2T} \left\{ F_T^*(\omega)\, G_T(\omega) \right\} = \lim_{T \to \infty} \frac{1}{2T} \left\{ F_T^*(\omega)\, F_T(\omega)\, H(\omega) \right\} \tag{10.28}$$

$$P_{fg}(\omega) = P_f(\omega)\, H(\omega). \tag{10.29}$$

Thus on Fourier transforming back to the time domain we have in the special case that $f(t)$ is white noise so that $P_f(\omega)$ is equal to a constant, say P, we get the required result:

$$R_{fg}(t) = P\, h(t). \tag{10.30}$$

This method of measuring $h(t)$ is attractive because it does not involve the high concentrations of input energy used in impulse testing and also because the test can be applied whilst the system is in use if some noise is added to the existing input and the correlation between the output and the injected noise is measured. If the noise is not white, then eqn (10.29) still holds and for instance noise which is white up to some cut-off frequency merely gives a slightly smoothed version of $h(t)$. In the general case eqn (10.30) is replaced by

$$R_{fg}(t) = h(t) \otimes R_f(t). \tag{10.31}$$

As a final comment on the correlator it is very useful to regard it as a filter whose transfer function can be adjusted at will. If we wish to simulate a filter with transfer function $H(\omega)$ and impulse response $h(t)$, then we feed the signal to be filtered $f(t)$ into one input of the correlator, and we feed a signal $h(-t)$ into the other. For an integration time constant long compared with the duration of $h(t)$ we obtain an output as follows for an input of finite duration:

$$g(t, \tau) = \int_{-\infty}^{+t} h(\tau - t_1)\, f(t_1)\, dt_1. \tag{10.32}$$

Provided the impulse response function satisfies the following requirement

$$h(\tau - t_1) = 0 \quad \text{for} \quad t_1 \geqslant t \tag{10.33}$$

we may rewrite eqn (10.32) as

$$g(t, \tau) = \int_{-\infty}^{+\infty} h(\tau - t_1)\, f(t_1)\, dt_1 \tag{10.34}$$

$$= h(\tau) \otimes f(\tau).$$

Regarded as a function of τ this is precisely the output required. If the impulse response is a causal function then eqn (10.33) will certainly be satisfied for $\tau < t$. It is interesting to note however that even if $h(t)$ is non-causal there is a possibility of satisfying eqn (10.33). Suppose $h(t)$ extends into negative time as far as some cut-off at $t = -t_c$. We now find that eqn (10.33) is satisfied provided

$$t \geqslant \tau + t_c. \tag{10.35}$$

In other words provided we are willing to wait long enough we can simulate a non-physical filter having a non-causal impulse response. There is nothing magical in this since our output is provided as a function of τ which is not the same as real time. Clearly if we are prepared to wait until the required duration of signal has arrived we can if we wish compute retrospectively the effect of any filter we like, so that the above procedure has not performed the impossible.

10.6 The Wiener–Hopf Condition

A common problem is the extraction of a random signal such as music or speech from a noisy or blurred waveform. The waveform may simply consist of the signal $s(t)$ plus noise $n(t)$, or for instance the signal may be blurred (i.e. convoluted) by the effects of echo or reverberation. A filter which suppresses frequencies present in the noise but passes those in the signal is a natural solution, but due to the randomness of the noise an accurate retrieval of the signal $s(t)$ may not be possible and the question arises as to what filter characteristic is optimum. The Wiener–Hopf condition given in eqn (10.36) below specifies the transfer function $H(\omega)$ of such an optimum filter in terms of the cross-power spectrum $P_{fs}(\omega)$ of $f(t)$ and $s(t)$ and of the power spectrum $P_f(\omega)$ of $f(t)$.

$$H(\omega) = \frac{P_{fs}(\omega)}{P_f(\omega)}. \tag{10.36}$$

Fourier inversion of this gives a relation involving the impulse response, $h(t)$:

$$R_f(t) \otimes h(t) = R_{fs}(t). \tag{10.37}$$

In the special case that the input $f(t)$ consists of signal plus noise so that

$$f(t) = s(t) + n(t) \tag{10.38}$$

eqns (10.37) and (10.36) may be simplified since in this case, assuming $s(t)$ and $n(t)$ are uncorrelated, we have

$$R_{fs}(t) = R_{(s+n)s}(t) = R_s(t) + R_{ns}(t) = R_s(t) \tag{10.39}$$

$$R_f(t) = R_{(s+n)}(t) = R_s(t) + R_n(t). \tag{10.40}$$

Equations (10.37) and (10.36) then become

$$\{R_s(t) + R_n(t)\} \otimes h(t) = R_s(t) \tag{10.41}$$

and

$$H(\omega) = \frac{P_s(\omega)}{P_s(\omega) + P_n(\omega)} . \tag{10.42}$$

This result is qualitatively as expected since such a filter passes frequency components present in $P_s(\omega)$ but suppresses those in $P_n(\omega)$. The impulse response derived from the Wiener–Hopf condition may not be causal in which case the filter is not physically realisable: this is however not so serious as it may seem since by multiplying $H(\omega)$ by a factor $e^{-i\omega t_0}$ we can perhaps delay $h(t)$ sufficiently for it to become physically realisable at least to a good approximation, the only effect being that the signal so recovered is delayed by an amount t_0.

The Wiener–Hopf condition given in eqns (10.36) and (10.37) covers situations more general than the simple burying of a signal in noise. For instance $f(t)$ may consist of the signal $s(t)$ blurred by reverberation so that

$$f(t) = s(t) \otimes m(t) \tag{10.43}$$

where $m(t)$ represents the reverberation. Using a subscript T to refer to quantities derived from a truncated version of the signal of duration T we may now write

$$P_{fs}(\omega) = \lim_{T \to \infty} \frac{1}{T} \left\{ F_T{}^*(\omega) \, S_T(\omega) \right\}$$

$$= \lim_{T \to \infty} \frac{1}{T} \left\{ S_T{}^*(\omega) \, M^*(\omega) \, S_T(\omega) \right\}$$

$$= P_s(\omega) \, M^*(\omega) \tag{10.44}$$

$$P_f(\omega) = \lim_{T \to \infty} \frac{1}{T} \left\{ [S_T{}^*(\omega) \, M^*(\omega)] \, [S_T(\omega) \, M(\omega)] \right\}$$

$$= P_s(\omega) \, M(\omega) \, M^*(\omega). \tag{10.45}$$

Thus the Wiener–Hopf condition of eqn (10.36) becomes

$$H(\omega) = \frac{1}{M(\omega)} . \tag{10.46}$$

This result is as expected since such an inverse characteristic is just what is required to counteract exactly the effect of the reverberation which was to multiply $F_T(\omega)$ by $M(\omega)$. Such a 'deconvoluting' filter may not be physically realisable due to $h(t)$ being non-causal, and due to the behaviour of $M(\omega)$ as $\omega \to \infty$ making its reciprocal non-transformable. However the second difficulty may be avoided by truncating $1/M(\omega)$ above some suitably high cut-off frequency, and the first difficulty may be overcome by delaying the impulse response suitably, so that an approximate deconvolution may be achieved.

The criterion which the Wiener–Hopf condition satisfies is that of minimum mean square error $\langle \varepsilon^2(t) \rangle$ where the error $\varepsilon(t)$ is defined as the difference between actual output $g(t)$ and desired output so that

$$\varepsilon(t) = g(t) - s(t). \tag{10.47}$$

Although the Wiener–Hopf filter may not be the optimum filter if non-linear systems are allowed, we can prove it to be the optimum linear filter as follows. Let the functions $H(\omega)$ and $h(t)$ satisfy eqns (10.36) and (10.37) and give rise to an output $g(t)$, for an input $f(t)$, with error $\varepsilon(t)$, whilst some other filter with characteristics $H_1(\omega)$ and $h_1(t)$ gives an output $g_1(t)$ with error $\varepsilon_1(t)$. Our proof consists in showing that $\langle \varepsilon_1^2(t) \rangle$ is greater than $\langle \varepsilon^2(t) \rangle$ irrespective of our choice of $H_1(\omega)$ and $h_1(t)$. First note that we may write

$$\langle \varepsilon_1^2(t) \rangle - \langle \varepsilon^2(t) \rangle$$

$$= \langle \{ \varepsilon_1(t) - \varepsilon(t) \}^2 \rangle + 2 \langle \varepsilon(t) \{ \varepsilon_1(t) - \varepsilon(t) \} \rangle$$

$$= \langle \{ g_1(t) - g(t) \}^2 \rangle + 2 \langle \varepsilon(t) \{ g_1(t) - g(t) \} \rangle. \tag{10.48}$$

Since the first average on the right hand side of eqn (10.48) is inherently either positive or zero, our result would be proven if we were to establish that the second average was zero. We will now establish that this is indeed the case by showing that $\varepsilon(t)$ and $\{ g_1(t) - g(t)$ are uncorrelated.

To establish this it is convenient to establish a general result applicable to any two finite power signals $x(t)$ and $y(t)$, and a finite energy signal $z(t)$. If we use the symbol $*$ to mean cross-correlation, as opposed to $\otimes$ meaning convolution, then

$$[x(t) * y(t)] \otimes z(t) = x(t) * [y(t) \otimes z(t)]. \tag{10.49}$$

This result can be verified either by writing out the implied integrations in full or by transforming to the frequency domain. Using the latter approach and using the subscript T to refer to quantities derived from truncated versions of $x(t)$ and $y(t)$ of duration T we may derive eqn (10.49) by Fourier inversion of the following obvious equality:

$$\lim_{T \to \infty} \frac{1}{T} \{[X_T(^*\omega) Y_T(\omega)] Z(\omega)\} = \lim_{T \to \infty} \frac{1}{T} \{X_T^*(\omega) [Y_T(\omega) Z(\omega)]\}. \quad (10.50)$$

We now establish that the cross correlation between $\varepsilon(t)$ and $\{g_1(t) - g(t)\}$ is zero.

$$[g_1(t) - g(t)] * [\varepsilon(t)] = [\{h_1(t) - h(t)\} \otimes f(t)] * [g(t) - s(t)] \quad (10.51)$$

$$= [h_1(t) - h(t)] \otimes [f(t) * \{g(t) - s(t)\}] \quad (10.52)$$

$$= [h_1(t) - h(t)] \otimes [f(t) * \{f(t) \otimes h(t) - s(t)\}] \quad (10.53)$$

$$= [h_1(t) - h(t)] \otimes [R_f(t) \otimes h(t) - R_{fs}(t)] \quad (10.54)$$

$$= 0. \quad (10.55)$$

Equation (10.54) collapses to zero on account of the Wiener–Hopf condition, eqn (10.37). Our result is thus proved.

The actual value of $\langle \varepsilon^2(t) \rangle$ may be derived as follows, working in the frequency domain and considering initially the case of truncated signals denoted by the subscript T.

$$\varepsilon_T(t) = g_T(t) - s_T(t) = f_T(t) \otimes h(t) - s_T(t). \quad (10.56)$$

The energy spectrum of $\varepsilon_T(t)$ is thus

$$|\{F_T(\omega) H(\omega) - S_T(\omega)\}|^2 = |F_T(\omega)|^2 |H(\omega)|^2 + |S_T(\omega)|^2$$
$$- F_T^*(\omega) H^*(\omega) S_T(\omega) - F_T(\omega) H(\omega) S_T^*(\omega). \quad (10.57)$$

On taking the limit as $T \to \infty$ and dividing through by T we now obtain

$$P_\varepsilon(\omega) = P_f(\omega) |H(\omega)|^2 + P_s(\omega) - P_{fs}(\omega) H^*(\omega) - P_{fs}^*(\omega) H(\omega) \quad (10.58)$$

$$= \frac{|P_{fs}(\omega)|^2}{P_f(\omega)} + P_s(\omega) - 2 \frac{|P_{fs}(\omega)|^2}{P_f(\omega)} \quad (10.59)$$

$$= P_s(\omega) - \frac{|P_{fs}(\omega)|^2}{P_f(\omega)}. \quad (10.60)$$

In deriving eqn (10.59) from (10.58) we have made use of the Wiener–Hopf condition, eqn (10.36). Finally use of eqn (4.10) gives us

$$\langle \varepsilon^2(t) \rangle = \frac{1}{2\pi} \int_{-\infty}^{+\infty} \left\{ P_s(\omega) - \frac{|P_{fs}(\omega)|^2}{P_f(\omega)} \right\} d\omega. \tag{10.61}$$

An alternative expression for $\varepsilon^2(t)$ may be obtained in the time domain as follows:

$$\langle \varepsilon^2(t) \rangle = \quad \langle \varepsilon(t) \{g(t) - s(t)\} \rangle \tag{10.62}$$

$$= - \langle \varepsilon(t) s(t) \rangle \tag{10.63}$$

$$= - \langle \{g(t) - s(t)\} s(t) \rangle \tag{10.64}$$

$$= - \langle \{f(t) \otimes h(t)\} s(t) \rangle + \langle s^2(t) \rangle \tag{10.65}$$

$$= - \left\langle \int_{-\infty}^{+\infty} h(t_1) f(t - t_1) s(t) dt_1 \right\rangle + \langle s^2(t) \rangle \tag{10.66}$$

$$= - \int_{-\infty}^{+\infty} h(t_1) \langle f(t - t_1) s(t) \rangle dt_1 + \langle s^2(t) \rangle \tag{10.67}$$

$$= - \int_{-\infty}^{+\infty} h(t_1) R_{fs}(t_1) dt_1 + \langle s^2(t) \rangle. \tag{10.68}$$

For the special case of signal plus noise considered in eqns (10.38)–(10.42) it is readily shown that eqn (10.61) reduces to the form

$$\langle \varepsilon^2(t) \rangle = \langle s^2(t) \rangle - \frac{1}{2\pi} \int_{-\infty}^{+\infty} \frac{\{P_s(\omega)\}^2}{P_s(\omega) + P_n(\omega)} d\omega. \tag{10.69}$$

10.7 The Matched Filter

We now consider the problem of an input waveform $f(t)$ consisting of a signal $s(t)$ of finite duration and known shape, but of unknown amplitude, buried in white noise $n(t)$. We wish to design a filter whose output $g(t)$ will rise to a large value when the signal arrives but which will be relatively small otherwise. Note in contrast to the situation discussed in Section 10.6 that we do not require the output to give a faithful reproduction of $s(t)$, rather we want to achieve a certain degree of bunching of the signal so as to have a large signal to noise ratio at the instant of maximum response. The optimum linear filter in this context is called the matched filter, because its characteristic is related uniquely to the particular signal being searched for. As we shall verify later the optimum filter is specified, apart from an arbitrary factor, by

$$H(\omega) = S^*(\omega) e^{-i\omega t_m} \qquad (10.70)$$

$$h(t) = s(t_m - t). \qquad (10.71)$$

In these equations t_m is equal to the time at which the response is maximum. Clearly if t_m is too early relative to the time when the signal arrives then we are effectively detecting the signal before it arrives, an impossibility reflected in the fact that $h(t)$ becomes non-causal. If t_m is made too large then although the system is realizable it gives a response which is needlessly delayed. In most cases t_m is best chosen to equal the time at which the arrival of the signal is just complete, which is in fact the least delay possible which is compatible with $h(t)$ being casual.

In problems of digital communication when one has to decide which out of a set vocabulary of signals is arriving one feeds the signal into a parallel array of filters matched to the various possibilities and sees which filter gives the largest output at the chosen time t_m. In other applications one may not know when a signal is due to arrive. One then merely waits for a large 'blip' to occur at the output.

One way of achieving a filter of given specification is to use a correlator as described in Section 10.5 (eqns (10.32)–(10.35)) with a signal $s(t - t_m)$ applied to the second input whilst the waveform to be processed is fed to the first input. The net result is that one is simply measuring a cross-correlation between $f(t)$ and $s(t - t_m)$. Clearly the output is a maximum for a delay t_m. One may figuratively say that the reference signal is searching for a copy of itself buried in the waveform $f(t)$.

Let us now see in what sense a matched filter is optimum. Let the output when white noise $n(t)$ alone is presented to the filter be $n_0(t)$, and let $s_0(t)$ be the output when the signal $s(t)$ is fed in by itself. We shall now prove that the matched filter defined by eqns (10.70) or (10.71) is that linear filter which gives a maximum value of $s_0{}^2(t_m)/\langle n_0{}^2(t) \rangle$. If $S_0(\omega)$ is the Fourier transform of $s_0(t)$ then

$$s_0(t) = \frac{1}{2\pi} \int_{-\infty}^{+\infty} S_0(\omega) e^{i\omega t} \, d\omega \qquad (10.72)$$

$$= \frac{1}{2\pi} \int_{-\infty}^{+\infty} S(\omega) H(\omega) e^{i\omega t} \, d\omega. \qquad (10.73)$$

Moreover from eqn (7.39) we may write $\langle n_0{}^2(t) \rangle$ as

$$\langle n_0{}^2(t) \rangle = \frac{1}{2\pi} \int_{-\infty}^{+\infty} P_n(\omega) |H(\omega)|^2 \, d\omega \qquad (10.74)$$

so that from eqns (10.73) and (10.74) we derive

$$\frac{s_0^2(t_m)}{\langle n_0^2(t)\rangle} = \frac{1}{2\pi}\frac{|\int_{-\infty}^{+\infty} S(\omega)\,H(\omega)\,e^{i\omega t_m}\,d\omega|^2}{\int_{-\infty}^{+\infty} P_n(\omega)\,|H(\omega)|^2\,d\omega}. \tag{10.75}$$

Now for complex functions $A(\omega)$ and $B(\omega)$ the Schwartz inequality (see Appendix F, eqns (F.2) and (F.7)) requires that

$$\left|\int_{-\infty}^{+\infty} A(\omega)\,B(\omega)\,d\omega\right|^2 \leqslant \int_{-\infty}^{+\infty} |A(\omega)|^2\,d\omega \int_{-\infty}^{+\infty} |B(\omega)|^2\,d\omega \tag{10.76}$$

the equality holding when

$$A(\omega) \propto B^*(\omega). \tag{10.77}$$

Thus putting

$$A(\omega) = S(\omega) \tag{10.78}$$

$$B(\omega) = H(\omega)\,e^{i\omega t_m} \tag{10.79}$$

we obtain

$$\frac{s_0^2(t_m)}{\langle n_0^2(t)\rangle} \leqslant \frac{\int_{-\infty}^{+\infty} |S(\omega)|^2\,d\omega \int_{-\infty}^{+\infty} |H(\omega)|^2\,d\omega}{2\pi\int_{-\infty}^{+\infty} P_n(\omega)\,|H(\omega)|^2\,d\omega}. \tag{10.80}$$

For the case of white noise we may put $P_n(\omega)$ equal to a constant, say P, over the frequency range of interest giving finally

$$\frac{s_0^2(t_m)}{\langle n_0^2(t)\rangle} \leqslant \frac{1}{2\pi P}\int_{-\infty}^{+\infty} |S(\omega)|^2\,d\omega \tag{10.81}$$

the optimum condition being when the equality of eqn (10.77) holds so that

$$S(\omega) \propto H^*(\omega)\,e^{-i\omega t_m}.$$

This is the result we wished to establish. For the optimum condition it is readily verified that eqn (10.81) may be written alternatively as

$$\frac{s_0^2(t_m)}{\langle n_0^2(t)\rangle} = \frac{\omega_c}{\pi}\frac{\int_{-\infty}^{+\infty} s^2(t)\,dt}{\langle n^2(t)\rangle} \tag{10.82}$$

where ω_c is a cut-off frequency for the white noise such that

$$\langle n^2(t)\rangle = \frac{1}{2\pi}\int_{-\infty}^{+\infty} P_n(\omega)\,d\omega = \frac{P\omega_c}{\pi}. \tag{10.83}$$

Chapter 11

Coherent Diffraction at Plane Apertures and Lenses

11.1 Introduction

When a plane wave is incident on a diffracting aperture the angular distribution of the emerging radiation (considered at large distances from the aperture) may be related to the spatial distribution of the amplitude and phase of the wave front when it emerges from the aperture. The relationship may elegantly be expressed in terms of a Fourier transform and in this chapter we start by establishing this. Two approaches are possible: one of these is based on Huygens' principle according to which each part of the wave front is considered as a source of secondary wave fronts whilst the other approach is based on the concept of the angular spectrum according to which the wave field is synthesised from a continuous distribution of plane progressive waves of fixed speed but differing directions. Each method requires care for a precise formulation and it is noteworthy that Huygens' principle has to be expressed in a special form arrived at by Fresnel and Kirchhoff.

Having discussed diffraction at a plane aperture using Fourier transform techniques we will see that the formulation can be generalised to cover the passage of a wave through optical systems and we will introduce an approach to optical theory based on the 'optical transfer function'. Such an approach will resemble closely that described in Chapter 7 for dealing with linear systems.

We will consider only scalar waves for which the disturbance is a scalar quantity. Although the diffraction of light (and all effects relating to polarization) is thus strictly excluded, it is nevertheless true that we shall uncover the basic principles of diffraction and that an extension of the

138

concepts to vector disturbances is best made after first understanding the scalar theory.

11.2 Co-ordinate System

We consider a plane wave disturbance advancing in the positive z direction towards an aperture in the $z = 0$ plane. The aperture modifies the phase and amplitude of the wave front and the disturbance continues onwards to give a diffracted wave pattern in the half space $z > 0$. Finite apertures are considered first, infinite ones being treated in the appropriate limit. The disturbance is considered as everywhere varying with time at an angular frequency ω and the free plane wave is characterised by a wavenumber $k = 2\pi/\lambda$ where λ is the wavelength. Polychromatic waves may be treated by a Fourier superposition of such monochromatic waves but we will not consider this aspect of the problem since it is not central to the phenomenon of diffraction.

Various methods of specifying the position of a point in the diffraction field will be useful in various contexts. We let $\mathbf{R}$ represent the position vector of such a point at (x, y, z) relative to an origin of co-ordinates chosen for convenience to lie somewhere within the aperture; equivalently the position may be specified by the pair $(\mathbf{S}, z)$ where $\mathbf{S}$ is a vector in the xy plane with components x and y (i.e. $\mathbf{S}$ is the projection of $\mathbf{R}$ on the xy plane). In certain contexts it is necessary to distinguish between a point A in the $z = 0$ plane (i.e. the aperture plane) and a point B somewhere in the diffraction field and we then represent the positions of A and B respectively by $(\mathbf{s}, 0)$ and $(\mathbf{S}, z)$. We let $\mathbf{r}$ represent a vector from A to B, so that $\mathbf{R} = \mathbf{s} + \mathbf{r}$, and we let θ be the angle between $\mathbf{r}$ and the z axis. The various quantities are illustrated in Fig. 11.1.

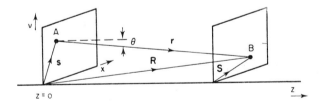

FIGURE 11.1

We will from time to time consider hypothetical plane waves traversing the radiation field and we let $\mathbf{k}$ represent the wave vector for such a disturbance with components k_x, k_y, k_z. Often it will be convenient to represent $\mathbf{k}$ by its projection $\mathbf{q}$ on the xy plane, so that $\mathbf{q}$ bears much the same relation to $\mathbf{k}$ as does S to R. Note that since the magnitude of **k** is fixed the wave vector is in fact completely specified by the vector **q** and we shall often use **q** instead of **k** for specifying a plane wave. **S** (or **s**) and **q** are both two-dimensional vectors and they will often appear as conjugate variables in a Fourier transformation. On occasion a two-dimensional vector **q** will be introduced simply as a vector conjugate to **S** (or **s**) without the specific interpretation just given being attached to it and we rely on the context to make the status of **q** clear in each case.

11.3 Treatment based on the Huygens, Fresnel, Kirchhoff Approach

Our aim is to derive the radiation field from a knowledge of the disturbance just after the aperture plane. Let the disturbance at a point B in the radiation field be $\phi(S, z, t)$. Since the oscillations are harmonic we may write $\phi(S, z, t)$ as follows and may construct a complex amplitude $f(S, z)$ to describe $\phi(S, z, t)$.

$$\phi(S, z, t) = A(S, z) \cos \{\omega t - \alpha(S, z)\} \tag{11.1}$$

$$= \mathrm{Re}\,[f(S, z)\,\mathrm{e}^{-i\omega t}] \tag{11.2}$$

where

$$f(S, z) = A(S, z) \exp\,[i\alpha(S, z)]. \tag{11.3}$$

The phase factor $\alpha(S, z)$ is a function of position but not of time, and this restriction defines the radiation field as *coherent*. As a special case the disturbance in the aperture plane may thus be written

$$\phi(s, 0, t) = \mathrm{Re}\,[f(s, 0)\,\mathrm{e}^{-i\omega t}] \tag{11.4}$$

and $f(s, 0)$ will be referred to as the aperture function.

In Appendix P we show, on the basis of a standard result worked out by Kirchhoff, that $f(S, z)$ and $f(s, 0)$ are related approximately as follows, with k, r and θ defined as in Section 11.2.

$$f(S, z) \approx -\frac{ik}{2\pi} \iint f(s, 0) \frac{\mathrm{e}^{ikr}}{r} \left(\frac{1 + \cos\theta}{2}\right) \mathrm{d}s. \tag{11.5}$$

The approximation holds provided $A(s, 0)$ and $\alpha(s, 0)$ change by only small fractional amounts over distances of the order of a wavelength (which implies that the aperture must be large compared with the wavelength) and provided also $r \gg \lambda$. These conditions are usually satisfied in optical problems, but not necessarily in microwave or acoustic diffraction. In these latter cases as we shall see the method based on the angular spectrum is superior. Equation (11.5) gives quantitative justification to Huygens' principle whereby the disturbance at B is considered as due to the cumulative effects of wavelets emitted by portions of the wave front in the aperture plane. The term e^{ikr}/r describes a spherical wave diverging from the element ds with an amplitude decaying inversely with distance, and therefore an intensity obeying an inverse square law. The term $(1 + \cos \theta)/2$ is an inclination factor not present in Huygens' simple ideas. The factor $-ik$ means that the re-emitted waves are a quarter of a period in advance of the original disturbance and that the strength of the emitted wavelet depends on wavelength. It is interesting historically that Fresnel preceeded Kirchhoff in seeing the need for the phase change and that he also introduced an inclination factor. However in Fresnel's theory the inclination factor was required simply to die away smoothly from unity to zero as θ changed from zero to $\pi/2$, no explicit expression being given. Kirchhoff's theory differs from Fresnel's in that the inclination factor still has the value of a half when $\theta = \pi/2$. Kirchhoff's treatment is based rigorously on a scalar wave equation and is described for instance in references 1, 4, 13, 17 and 18.

If in addition to the assumptions made above the dimensions of the aperture are small compared with r then we have what is known as the Fraunhofer approximation and we may write, as some simple geometry shows,

$$kr \approx kR - \frac{k}{R} \mathbf{s} \cdot \mathbf{S}.$$

In this approximation eqn (11.5) now becomes

$$f(\mathbf{S}, z) \approx - \frac{ik}{2\pi R} \left(\frac{1 + \cos \theta}{2} \right) e^{ikR} \iint \exp \left(- \frac{ik}{R} \mathbf{s} \cdot \mathbf{S} \right) f(\mathbf{s}, 0) \, ds. \quad (11.6)$$

This relation has the appearance of a Fourier transformation, and it is convenient now to define a function $F(\mathbf{q}, z)$ as the two-dimensional transform of $f(\mathbf{S}, z)$ according to

$$F(\mathbf{q}, z) = \iint \exp \left(-i \, \mathbf{q} \cdot \mathbf{S} \right) f(\mathbf{S}, z) \, d\mathbf{S}. \quad (11.7)$$

We now write equation 11.6 as

$$f(\mathbf{S}, z) \approx - \frac{ik \, e^{ikR}}{2\pi R} \frac{(1 + \cos \theta)}{2} F\left(\frac{k}{R} \mathbf{S}, 0 \right). \quad (11.8)$$

Note that the quantity $k\mathbf{S}/R$ depends on the direction of a line from the origin to the point B but not on the distance, so that the term involving the function F in eqn (11.8) is a purely directional term and is the one displaying the diffractive effects. We now drop the factor $(1 + \cos \theta)/2$ since our starting eqn (11.5) is only valid if $f(\mathbf{s}, 0)$ varies sufficiently slowly with $\mathbf{s}$, which means that $F(\mathbf{q}, 0)$ is concentrated within the region $q \lesssim 1/\lambda$ which in turn means that eqn (11.8) will only be valid when most of the radiation is diffracted into forward directions for which $\cos \theta \approx 1$. Thus we write as our final expression for Fraunhofer diffraction

$$f(\mathbf{S}, z) \approx - \frac{ik \, e^{ikR}}{2\pi R} F\left(\frac{k}{R} \mathbf{S}, 0\right). \tag{11.9}$$

We may express this result in words by saying that the angular distribution of diffracted amplitude is proportional to the Fourier transform of the aperture function.

11.4   Treatment based on the Angular Spectrum

Our aim as in the previous section is to derive the radiation field from a knowledge of the aperture function and thus derive $f(\mathbf{S}, z)$ from $f(\mathbf{s}, 0)$. We now proceed from the physical assumption (justifiable in terms of a scalar wave equation) that the diffraction field may be synthesised as a superposition of plane wave disturbances of fixed wavelength but of various amplitudes and travelling in various directions. Noting that the wave vector $\mathbf{k}$ of such a wave is specified uniquely by its projection $\mathbf{q}$ on the xy plane we now express this synthesis as follows

$$\phi(\mathbf{R}, t) = \iint_{q < k} B(\mathbf{q}) \cos \{\mathbf{k} \cdot \mathbf{R} - \omega t + \beta(\mathbf{q})\} \, d\mathbf{q} \tag{11.10}$$

$$= \mathrm{Re} \left[e^{-i\omega t} \iint_{q < k} G(\mathbf{q}) \exp (i\mathbf{k} \cdot \mathbf{R}) \, d\mathbf{q} \right] \tag{11.11}$$

where

$$G(\mathbf{q}) = B(\mathbf{q}) \exp [i\beta(\mathbf{q})]. \tag{11.12}$$

The amplitude and phase of a component wave are thus described by the functions $B(\mathbf{q})$ and $\beta(\mathbf{q})$, and the resulting complex function $G(\mathbf{q})$ is called the complex angular spectrum of the array of plane waves. The integrations in (11.10) and (11.11) are intended to cover all values of the vector $\mathbf{q}$ terminating within a circle of radius k since our definition of $\mathbf{q}$ does not yet allow meaning to be given to values of q greater than k.

Comparison of eqns (11.11) and (11.2), bearing in mind that each is valid for all values of t, leads to the equality

$$f(\mathbf{S}, z) = \iint_{q<k} G(\mathbf{q}) \exp(i\mathbf{k} \cdot \mathbf{R}) \, d\mathbf{q} \qquad (11.13)$$

$$= \iint_{q<k} G(\mathbf{q}) \exp(ik_z z) \exp(i\mathbf{q} \cdot \mathbf{S}) \, d\mathbf{q}. \qquad (11.14)$$

As a special case of eqn (11.14) we have

$$f(\mathbf{s}, 0) = \iint_{q<k} G(\mathbf{q}) \exp(i\mathbf{q} \cdot \mathbf{s}) \, d\mathbf{q}. \qquad (11.15)$$

Equation (11.15) is strongly reminiscent of a Fourier transform and is only spoilt by the finite limit on the region of integration. As we shall see in the next section a useful meaning can be attached to values of q greater than k and the limits can then be extended to make the relation a true Fourier transform. However we can already see a method of solving the problem we set ourselves in the case that $G(\mathbf{q})$ is confined to values of q much less than k (i.e. the angular spectrum is narrow angle) so that the limits in eqns (11.14) and (11.15) may be extended to infinity formally without loss of accuracy. Given $f(\mathbf{s}, 0)$ we use the inversion of (11.15) to find $G(\mathbf{q})$ and then use (11.14) to obtain $f(\mathbf{S}, z)$ by a subsequent transformation applied to $\exp(ik_z z) G(\mathbf{q})$. In an intuitive way we may further guess that the complex amplitude at a point a long way from a small diffracting aperture will be proportional to the amplitude of the component plane wave which is directed towards the point in question. This guess tallies qualitatively with eqn (11.9) which was derived by a different method. We proceed more rigorously in Section 11.6.

Before proceeding further with the development however, let us see how it is useful to attach meaning to values of q greater than k.

11.5 Evanescent Waves

An evanescent wave is the name given to a disturbance described by

$$\phi(x, y, z, t) \propto e^{-\mu z} \cos\{k_x x + k_y y - \omega t + \alpha\} \qquad (11.16)$$

with

$$k_x^2 + k_y^2 - \mu^2 = k^2 = \frac{\omega^2}{c^2}. \qquad (11.17)$$

It represents a harmonic disturbance travelling over the $z = 0$ plane at a speed slower than the wave velocity c and wavelength less than that of free travelling waves. In the z direction the amplitude is attenuated exponentially. It is easily verified that such a wave satisfies the scalar wave equation

$$\nabla^2 \phi = \frac{1}{c^2} \frac{\partial^2 \phi}{\partial t^2}.$$ (11.18)

If in this context we define $\mathbf{q}$ as a vector in the xy plane with components k_x and k_y then the resultant of a distribution of such waves with various $\mathbf{q}$ values, amplitudes and phases may be written

$$\phi(\mathbf{S}, z, t) = \iint_{q>k} e^{-\mu z} B(\mathbf{q}) \cos \{\mathbf{q} \cdot \mathbf{S} - \omega t + \alpha(\mathbf{q})\} \, d\mathbf{q}$$ (11.19)

$$= \mathrm{Re}' \left[e^{-i\omega t} \iint_{q>k} G(\mathbf{q}) \, e^{-\mu z} \exp(i\mathbf{q} \cdot \mathbf{S}) \, d\mathbf{q} \right]$$ (11.20)

where

$$G(\mathbf{q}) = B(\mathbf{q}) \exp[i\alpha(\mathbf{q})].$$ (11.21)

The functions $B(\mathbf{q})$, $\alpha(\mathbf{q})$ and $G(\mathbf{q})$ now refer to the spectrum of evanescent waves, and the integrations cover all values of $\mathbf{q}$ with $q > k$, but otherwise eqns (11.19)–(11.21) are formally identical to eqns (11.10)–(11.12) if we allow k_z to become complex and write it as $i\mu$. Equations (11.14) and (11.15) receive as their analogues the following:

$$f(\mathbf{S}, z) = \iint_{q>k} G(\mathbf{q}) \, e^{-\mu z} \exp(i\mathbf{q} \cdot \mathbf{S}) \, d\mathbf{q}$$ (11.22)

and

$$f(\mathbf{s}, 0) = \iint_{q>k} G(\mathbf{q}) \exp(i\mathbf{q} \cdot \mathbf{s}) \, d\mathbf{q}$$ (11.23)

so that provided the symbols receive the correct interpretations in the appropriate domains, we can now combine the equations of this section with those of the previous section to express the combined results of a spectrum of plane waves and of evanescent waves to give finally

$$f(\mathbf{S}, z) = \iint G(\mathbf{q}) \exp(i k_z z) \exp(i\mathbf{q} \cdot \mathbf{S}) \, d\mathbf{q}$$ (11.24)

$$f(\mathbf{s}, 0) = \iint G(\mathbf{q}) \exp(i\mathbf{q} \cdot \mathbf{s}) \, d\mathbf{q}.$$ (11.25)

These expressions now represent true Fourier type transforms since the limits of integration cover all values of $\mathbf{q}$, and the expressions may if required be inverted. We have in fact already defined $F(\mathbf{q}, z)$ formally as the Fourier transform of $f(\mathbf{S}, z)$ in equation 11.7 and we see that $F(\mathbf{q}, z)$ may be related to the angular spectrum as follows:

$$F(\mathbf{q}, z) = (2\pi)^2 \exp(ik_z z) \, G(\mathbf{q}) \qquad (11.26)$$

where

$$k_z = \sqrt{k^2 - q^2}. \qquad (11.27)$$

For the special case that $z = 0$ we see that apart from the trivial factor of $(2\pi)^2$ the angular spectrum is equal to the Fourier transform of the aperture function. This of course explains the well known fact that narrow apertures lead to broad diffraction patterns. It is also worth noting that evanescent waves only become necessary in the synthesis when the aperture contains structure finer than the wavelength so that the diffraction is wide angle and is 'trying' to extend beyond 90°. Since evanescent waves die away quickly in the z direction this means that the diffraction pattern at large values of z depends only on the coarse structure in the aperture.

11.6   Summary of Results using the Two Methods

The Kirchhoff approach has given us two approximate formulae (eqns (11.5) and (11.9)) relating $f(\mathbf{S}, z)$ to $f(\mathbf{s}, 0)$. The method of the angular spectrum has given us an exact result in which $f(\mathbf{S}, z)$ may be related to $f(\mathbf{s}, 0)$ by using eqn (11.26) to relate $F(\mathbf{q}, z)$ to $F(\mathbf{q}, 0)$ as in eqn (11.28) below. This relation between the transforms is not readily transformed to give a direct relation between $f(\mathbf{S}, z)$ and $f(\mathbf{s}, 0)$, but it is instructive to make some successive approximations and see how results emerge that are similar to those based on the Kirchhoff approach. The first approximation is to assume that $F(\mathbf{q}, 0)$ can be approximated to zero outside a range of $\mathbf{q}$ values defined by $q \ll k$. This is equivalent to assuming that the diffraction is forward directed and evanescent waves are absent. The second approximation is to assume in addition that $f(\mathbf{s}, 0)$ can be approximated to zero outside a range of s values defined by $s \ll z$. This is equivalent to considering only points at distances far away compared with the aperture size: this is the Fraunhofer region.

Below we summarise the results obtained, starting with eqn (11.28) which follows directly from eqn (11.26). The further derivations are described in Appendix Q.

Exact. $$F(\mathbf{q}, z) = \exp(iz\sqrt{k^2 - q^2}) \, F(\mathbf{q}, 0). \qquad (11.28)$$

Exact. $f(S, z) = f(S, 0) \otimes \left[\frac{1}{(2\pi)^2} \iint \exp(i\mathbf{q} \cdot \mathbf{S}) \{\exp(iz\sqrt{k^2 - q^2})\} \, d\mathbf{q} \right].$

$$(11.29)$$

Fresnel. $F(\mathbf{q}, z) = e^{ikz} \{\exp(-izq^2/2k) F(\mathbf{q}, 0)\}.$ (11.30)

Fresnel. $f(S, z) = -\frac{ik}{2\pi z} e^{ikz} \{f(S, 0) \otimes \exp(ikS^2/2z)\}$ (11.31)

$$= -\frac{ik}{2\pi z} \exp[ik(z + S^2/2z)] \iint \exp\left(-i\frac{k}{z}\mathbf{S} \cdot \mathbf{s}\right)$$

$$\times \{f(\mathbf{s}, 0) \exp(iks^2/2z)\} \, d\mathbf{s}. \qquad (11.32)$$

Fraunhofer. $f(S, z) = -\frac{ik}{2\pi z} \iint f(\mathbf{s}, 0) e^{ikr} \, d\mathbf{s}$ (11.33)

$$= -\frac{ik}{2\pi z} e^{ik\,R} \iint f(\mathbf{s}, 0) \exp\left(-i\frac{k}{z}\mathbf{s} \cdot \mathbf{S}\right) d\mathbf{s} \quad (11.34)$$

$$= -\frac{ik}{2\pi z} e^{ikR} F\left(\frac{k}{z}\mathbf{S}, 0\right). \qquad (11.35)$$

Let us interpret these results. Equation (11.28) shows that the angular spectrum of the radiation field remains the same with increasing z except for variations in phase of the various angular components. The energy spectrum $|F(\mathbf{q}, z)|^2$ is thus independent of z. Turning attention to eqn (11.29) we see already some similarity with a Huygens' type theory since the convolution means that the amplitude at each point in the $z = 0$ plane becomes 'spread out' by the expression in the square bracket as increasing values of z are considered. This spread function is involved and it is only after various approximations that we reach eqn (11.33) which is a direct expression of Huygens' ideas. It must be mentioned here that we have rather oversimplified our statement of the approximations involved; we discuss the approximations more fully in Appendix Q.

It will be noticed that eqn (11.32) makes use of a transform of the type

$$\int e^{ixy} f(x) e^{-iax^2} \, dx. \qquad (11.36)$$

Such a transform is known as a Fresnel transform. For the special case that $f(x)$ is unity in the range $0 \leqslant x \leqslant X$ and equal to zero otherwise the integral is known as a Fresnel integral. With a put equal to $\pi/2$ the real and imaginary parts of a Fresnel integral are commonly written as $C(X)$ and $S(X)$ and are tabulated for instance in references 8 and 18. The values of $C(X)$ and $S(X)$ plotted against each other form what is known as Cornu's spiral, the basis of a convenient graphical method of dealing with simple apertures.

11.7   Power Flow and Intensity

For most wave phenomena the intensity is proportional to the square of the wave amplitude, and thus to $|f(\mathbf{R})|^2$. It is satisfactory that we may use Parseval's theorem to verify that the total energy flow across the aperture plane equals that across any subsequent plane. Making use of eqns (3.16) and (11.28) we have:

$$\iint |f(\mathbf{S}, z)|^2 \, d\mathbf{S} = \frac{1}{(2\pi)^2} \iint |F(q, z)|^2 \, d\mathbf{q}$$

$$= \frac{1}{(2\pi)^2} \iint |F(q, 0)|^2 \, d\mathbf{q}$$

$$= \iint |f(s, 0)|^2 \, d\mathbf{s}.$$

If we consider the aperture function $f(\mathbf{s}, 0)$ as resulting from the passage of an incident plane wave of unit amplitude through a diffracting screen of some sort then in the Fraunhofer approximation it is convenient to express the angular distribution of diffracted radiation by the ratio of the energy diffracted into a solid angle $d\Omega$ in some direction to the incident energy per unit area. Writing this ratio as $\sigma \, d\Omega$, where σ (the differential cross-section) depends on the direction of scattering, we have

$$\sigma \, d\Omega = |f(\mathbf{S}, z)|^2 \, dS \tag{11.37}$$

where the area dS on the plane at z subtends an angle $d\Omega$ at the diffracting aperture. Within the forward angle approximation being used we have $dS/d\Omega \approx R^2 \approx z^2$ so that using eqn (11.35) we obtain

$$\sigma = z^2 |f(\mathbf{S}, z)|^2 = \frac{k^2}{(2\pi)^2} |F(\mathbf{q}, 0)|^2$$

$$= \frac{k^2}{(2\pi)^2} S_f(\mathbf{q}, 0) \quad \left[\mathbf{q} = \frac{k}{R} \mathbf{S} \right]. \tag{11.38}$$

If the aperture function is a finite power rather than a finite energy function, in the sense used in eqn (4.22), then the quantity σ can tend to infinity and it is convenient to use instead the ratio of the diffracted energy per steradian *per unit area of aperture* to the incident energy per unit area. Using $I \, d\Omega$ to represent this ratio we get

$$I = \lim_{S \to \infty} \frac{k^2}{(2\pi)^2} \frac{1}{S} |F_S(\mathbf{q}, 0)|^2 = \frac{k^2}{(2\pi)^2} P_f(\mathbf{q}, 0) \tag{11.39}$$

where $F_S(\mathbf{q}, 0)$ is obtained from a truncated version of $f(\mathbf{s}, 0)$ which is equal to zero outside of some area S.

11.8 Diffraction by One-Dimensional Apertures

To describe the diffraction of waves moving in a two-dimensional space at a one-dimensional aperture the previous results are easily adapted by replacing the two-dimensional transforms by one-dimensional ones. Using x and X instead of $\mathbf{s}$ and $\mathbf{S}$ to refer to distance measured perpendicular to the z axis, and replacing $\mathbf{q}$ by the scalar quantity q, then eqns (11.28)–(11.35) and (11.38)–(11.39) become replaced by:

Exact. $$F(q, z) = \exp\left(iz\sqrt{k^2 - q^2}\right) F(q, 0). \tag{11.40}$$

Exact. $$f(X, z) = f(X, 0) \otimes \left[\frac{1}{2\pi} \int e^{iqX} \{\exp\left(iz\sqrt{k^2 - q^2}\right)\} \, dq \right]. \tag{11.41}$$

Fresnel. $$F(q, z) = e^{ikz} \{\exp\left(-izq^2/2k\right) F(q, 0)\}. \tag{11.42}$$

Fresnel. $$f(X, z) = \frac{(1 - i)}{\sqrt{2}} \sqrt{\frac{k}{2\pi z}} e^{ikz} \{f(X, 0) \otimes \exp\left(ikX^2/2z\right)\} \tag{11.43}$$

$$= \frac{(1 - i)}{\sqrt{2}} \sqrt{\frac{k}{2\pi z}} \exp\left[ik(z + X^2/2z)\right] \int \exp\left(-i\frac{k}{z} Xx\right)$$
$$\times \{f(x, 0) \exp\left(ikx^2/2z\right)\} \, dx. \tag{11.44}$$

Fraunhofer. $$f(X, z) = \frac{(1 - i)}{\sqrt{2}} \sqrt{\frac{k}{2\pi z}} \int f(x, 0) \, e^{ikr} \, dx \tag{11.45}$$

$$= \frac{(1 - i)}{\sqrt{2}} \sqrt{\frac{k}{2\pi z}} e^{ikR} F\left(\frac{k}{z} X, 0\right). \tag{11.46}$$

$$\sigma = \frac{k}{2\pi} |F(q, 0)|^2 = \frac{k}{2\pi} S_f(q, 0) \left[q \approx \frac{k}{z} X \approx k\theta\right]. \tag{11.47}$$

$$I = \lim_{L \to \infty} \frac{k}{2\pi} \frac{1}{L} |F_L(q, 0)|^2 = \frac{k}{2\pi} P_f(q, 0). \tag{11.48}$$

The factor $(1 - i)/\sqrt{2}$ occurs, representing a phase advance of $\pi/4$, due to the use of transform 2.88 instead of its two-dimensional equivalent, eqn (3.25a).

It is tempting to wonder whether diffraction by three-dimensional objects as in X-ray and neutron diffraction can be treated simply by changing to three-dimensional transforms. This is not so however, since the result would be applicable to waves moving in a four-dimensional space. The appropriate formalism for X-ray or neutron diffraction is established in Chapter 13.

11.9   Examples of One-Dimensional Systems. Babinet's Principle

Although practicable diffraction apertures are necessarily two-dimensional they may often be treated using the one-dimensional formalism. This is so when $f(s, 0)$ may be written as a product $f_x(x, 0) f_y(y, 0)$ so that $F(q, 0)$ may be written $F_x(q_x, 0) F_y(q_y, 0)$ and when the dominant diffraction effects concern only one or other dimension. This is so for instance with parallel arrays of long slits as in the diffraction grating.

The angular distributions of diffracted energy for many simple systems may be derived directly from the table of transforms in Chapter 2 using eqns (11.47) and (11.48). We give here a few examples by way of illustration. For a *slit of width* 2L we get from (2.48) after putting $y = q \approx k\theta \approx 2\pi\theta/\lambda$,

$$\sigma = \frac{4L^2}{\lambda} \left(\frac{\sin kL\theta}{kL\theta} \right)^2 . \tag{11.49}$$

From eqn (2.49) it is evident that shifting the slit does not affect the angular distribution. The *double slit* follows from eqn (2.50).

The convolution theorem (Chapter 5) provides the basis for a result well known experimentally. If the aperture function is the convolution of two functions then the angular distribution is the product of the individual distributions (see eqn (4.31)). For instance if an arrangement of narrow slits gives one pattern the effect of widening the apertures is merely to multiply the overall intensity distribution by a factor corresponding to the intensity distribution for one aperture. Thus for *N slits of separation* x_0 and each of width 2L we get from eqns (2.48), (2.71) and (11.47):

$$\sigma = \frac{4L^2}{\lambda} \left\{ \frac{\sin (Nx_0 k\theta/2)}{\sin (x_0 k\theta/2)} \right\}^2 \left\{ \frac{\sin (Lk\theta)}{Lk\theta} \right\}^2 . \tag{11.50}$$

An *amplitude modulated aperture* in which the amplitude varies harmonically across a diffracting aperture of finite width is treated using transforms 2.52 or 2.53. If the harmonic modulation appears superimposed on a steady amplitude (as is likely in practice) then we must add the transform of eqn (2.48) before squaring the modulus. The result is complicated, but can be seen to give two diffracted maxima together with a central peak the widths of the peaks decreasing as the aperture size increases. If the period x_0 of the modulation is small compared with the overall width of the aperture we obtain the approximation that if

$$f(x, 0) = A + B \cos (2\pi x/x_0) \qquad [|x| \leqslant L]$$

$$= 0 \qquad\qquad\qquad [|x| > L] \qquad (11.51)$$

then

$$\sigma \approx \frac{4L^2}{\lambda} \left[A^2 \left\{ \frac{\sin Lk\theta}{Lk\theta} \right\}^2 + \frac{B^2}{4} \left\{ \frac{\sin L(k\theta - 2\pi/x_0)}{L(k\theta - 2\pi/x_0)} \right\}^2 \right.$$

$$\left. + \frac{B^2}{4} \left\{ \frac{\sin L(k\theta + 2\pi/x_0)}{L(k\theta + 2\pi/x_0)} \right\}^2 \right]. \qquad (11.52)$$

A *diffracting screen that is infinitely wide* is of course not physically realizable but is a convenient abstraction for describing the important features of apertures that are very wide compared with the dimensions of fine structure which they may contain. We now utilise the power spectrum of the aperture function as is implied in eqn (11.48). If the aperture function is random in some respect then we use the methods of Chapter 6 to obtain the power spectrum. If the aperture function $f(x, 0)$ is repetitive in some way so that $F(q, 0)$ defined according to

$$F(q, 0) = \int e^{-iqx} f(x, 0) \, dx \qquad (11.53)$$

contains delta functions then eqn (4.11) is useful in evaluating eqn (11.48) so that if

$$F(q, 0) = \sum_n a_n \delta(q - q_n) \qquad (11.54)$$

then

$$I = \frac{k}{(2\pi)^2} \sum_n |a_n|^2 \delta(q - q_n) \qquad (11.55)$$

$$= \frac{1}{(2\pi)^2} \sum_n |a_n|^2 \delta(\theta - q_n/k). \qquad (11.56)$$

As an example an amplitude modulated grating of infinite extent for which

$$f(x, 0) = A + B \cos (2\pi x/x_0) \qquad [-\infty < x < +\infty] \qquad (11.57)$$

is treated using eqns (2.75) and (2.59) so that

$$F(q, 0) = 2\pi A \delta(q) + \pi B \, \delta(q - 2\pi/x_0) + \pi B \, \delta(q + 2\pi x_0) \qquad (11.58)$$

and

$$I = A^2 \, \delta(\theta) + \frac{B^2}{4} \{\delta(\theta - \lambda/x_0) + \delta(\theta + \lambda/x_0)\} \qquad (11.59)$$

A *refracting wedge*, or thin prism, of infinite extent which has the effect of altering the phase of the wavefront by an amount proportional to x (say αx) is treated using eqn (2.58) and gives

$$I = \delta(\theta - \alpha\lambda/2\pi). \qquad (11.60)$$

This corresponds to a sharp but deviated 'ray' as expected for a prism.

A *thin sheet of refracting material of oscillating thickness* will give an aperture function of constant amplitude but oscillating phase. If the oscillations are periodic with period x_0 then

$$f(x, 0) = \exp [i \, a \cos (2\pi x/x_0)] \qquad (11.61)$$

and from eqn (2.80) we get

$$I = \sum_{n=-\infty}^{+\infty} \{J_n(a)\}^2 \, \delta(\theta - n\lambda/x_0). \qquad (11.62)$$

An *infinite grating of narrow slits* can be described by an aperture function

$$f(x, 0) = f_0(x) \otimes g(x)$$

where $f_0(x)$ consists of a series of delta functions defining the positions of the slits and $g(x)$ is the characteristic aperture function for one slit. If there is a regularity in the arrangement of the slits then $F_0(q)$ will consist of delta functions and we may write

$$F(q, 0) = F_0(q) \, G(q) \qquad (11.63)$$

$$= \text{say}, \qquad G(q) \sum_n a_n \, \delta(q - q_n). \qquad (11.64)$$

It then follows from eqn (11.56) that

$$I = \frac{1}{(2\pi)^2} |G(k\theta)|^2 \sum_n |a_n|^2 \delta(\theta - q_n/k).$$

(11.65)

For instance eqn (2.72) shows that *regularly spaced slits a distance x_0 apart* give sharp peaks at $\theta = \pm n\lambda/x_0$ with $n = 0, 1, 2, \ldots$. Equations (2.74) and (2.89) show that periodic 'errors' in the amplitudes or positions of the slits give rise to 'satellites' or 'ghosts' on either side of each main diffraction peak.

Let us consider now the effect of placing an object in the path of an infinite wave front. As an example consider an *opaque object of width $2L$* so that the aperture function is

$$f(x, 0) = 0 \quad [|x| \leqslant L]$$
$$= 1 \quad [|x| > L]$$

(11.66)

and so from eqn (2.79)

$$F(q, 0) = 2\pi\delta(q) - \frac{2 \sin Lq}{q}.$$

(11.67)

If we regard the delta function in (11.69) as representing simply the continuation forwards of the incident wavefront, then the diffracted radiation is accounted for by ignoring this delta function. We now use eqn (11.47) to give

$$\sigma = \frac{4L^2}{\lambda} \left(\frac{\sin Lk\theta}{Lk\theta} \right)^2 \quad [\theta \neq 0].$$

(11.68)

This expression integrated over all angles comes to equal $2L$, showing that the total diffracted energy equals that incident on the width of the object: this result is not so obvious as it may seem as the next examples will show.

Consider now a *refracting 'slab' of width $2L$* which alters the phase of the wavefront locally by an amount ϕ so that the aperture function is

$$f(x, 0) = 1 \quad [|x| > L]$$
$$= e^{i\phi} \quad [|x| \leqslant L].$$

(11.69)

From eqns (2.79) and (2.48) we obtain

$$F(q, 0) = 2\pi\delta(q) - \frac{2 \sin Lq}{q} + 2 e^{i\phi} \frac{\sin Lq}{q}.$$

(11.70)

Accordingly the cross-section is given by eqn (11.47) as

$$\sigma = \frac{8L^2 (1 - \cos \theta)}{\lambda} \left\{ \frac{\sin Lk\theta}{Lk\theta} \right\}^2 \quad [\theta \neq 0] \qquad (11.71)$$

This shows that the angular dependence of the diffraction is the same as for the opaque object, but that the intensity is different by a factor which depends on ϕ. Clearly if ϕ is a multiple of 2π the resulting wavefront $f(x, 0)$ is uniform so that the slab might just as well not be there and the intensity is zero as indicated by eqn (11.71). However if ϕ is equal to π the intensity rises to four times that produced by the opaque object. It is perhaps surprising that a transparent object can diffract more strongly then an opaque one of the same size: the explanation lies in the fact that whereas the opaque object reduces the amplitude of the wave front to zero locally the transparent object can actually reverse the sign of the wave amplitude (corresponding to a phase change of π) so that the overall disturbance to the wave front is greater. An alternative way of explaining how the cross-section of the object can be up to four times its geometrical size is made possible by the next example.

Consider now a *refracting wedge or prism of width* $2L$ which modifies the phase of the wave front so that

$$f(x, 0) = 1 \qquad\qquad [|x| > L]$$

$$= e^{i(\alpha x + \beta)} \qquad [|x| \leq L]. \qquad (11.72)$$

From eqns (2.79) and (2.51) we get

$$F(q, 0) = 2\pi\delta(q) - \frac{2 \sin Lq}{q} + e^{i\beta} \frac{2 \sin \{L(\alpha - q)\}}{(\alpha - q)} . \qquad (11.73)$$

If $\alpha L \gg 1$ the two sine terms give negligible overlap and on squaring the modulus according to eqn (11.47) we obtain

$$\sigma = \frac{4L^2}{\lambda} \left(\frac{\sin Lk\theta}{Lk\theta} \right)^2 + \frac{4L^2}{\lambda} \left[\frac{\sin \{L(\alpha - k\theta)\}}{L(\alpha - k\theta)} \right] \quad [\theta \neq 0]. \qquad (11.74)$$

In this case the total diffracted energy integrated over all angles is equal to twice the energy incident on the width $2L$, the energy being divided equally between forward and refracted peaks each of width λ/L. We may regard the forward peak as appropriate to an infinite wave front with a portion of width $2L$ missing, and the deviated peak at angle $\theta = \alpha/k$ as due to the short length of wave front which emerges from the wedge. So long as the peaks are well separated no interference between them occurs; however if $\alpha L \lesssim 1$ the peaks begin to overlap and one must consider interference effects which allow either desstructive subtraction or addition of amplitudes so that the intensity can lie between zero and four times the intensity of either peak by itself. The previous example leading to eqn (11.71) can be considered as the limiting case when $\alpha \to 0$ and both peaks are superposed constructively.

A comparison of eqns (11.49) and (11.68) shows that the diffraction patterns of a hole in a screen and of an opaque object the same size as the hole are the same, apart from the forward delta function. This is an example illustrating *Babinet's principle*, which relates the diffraction patterns of complementary diffracting screens. Complementary diffracting screens are ones for which the individual aperture functions add up to unity for all values of x. For instance a system of holes in an opaque screen is complementary to a system of opaque objects having the same sizes and positions as the holes. If subscripts 1 and 2 refer to a pair of such screens then we may prove the principle as follows:

$$f_1(x, 0) + f_2(x, 0) = 1 \qquad (11.75)$$

$$F_1(q, 0) + F_2(q, 0) = 2\pi \, \delta(q) \qquad (11.76)$$

so that

$$|F_1(q, 0)|^2 = |F_2(q, 0)|^2 \qquad [q \neq 0]. \qquad (11.77)$$

A modified form of Babinet's principle is often useful in which we consider complementary screens to be placed in turn across some aperture of finite size. In this situation the diffraction patterns are only equal at those angles where in the absence of either screen but with the aperture still present the intensity would be zero. The necessary modification of the above proof is very straightforward.

11.10   Examples of Two-Dimensional Systems

The two-dimensional transforms given in Section 3.7 may be used directly to give the intensity distributions for many common systems if we make use of eqns (11.38) and (11.39).

A *circular hole of radius R* follows from eqn (3.27) giving

$$\sigma = k^2 R^4 \left\{ \frac{J_1(Rk\theta)}{Rk\theta} \right\}^2. \qquad (11.78)$$

As shown in Appendix H this intensity pattern consists of a bright central spot surrounded by rings, the first minimum in intensity occurring when

$$\theta = 3{\cdot}832/Rk = 1{\cdot}22\lambda/2R. \tag{11.79}$$

A *rectangular hole in a screen* is covered by eqn (3.26) giving

$$\sigma = \frac{16a^2\, b^2}{\lambda^2} \left\{ \frac{\sin\,(ak\theta_x)}{ak\theta_x}\,\frac{\sin\,(bk\theta_y)}{bk\theta_y} \right\}^2. \tag{11.80}$$

In eqn (11.80), $2a$ represents the width of the aperture in the x-direction, and $2b$ is the width in the y-direction. θ_x and θ_y represent angles between the diffracted 'ray-direction' and the yz and xz planes respectively, so that the components of the vector $\mathbf{q}$ (eqn (11.38)) are given in the forward angle approximation by:

$$q_x = \frac{k}{R}\,X \approx k\theta_x$$

$$q_y = \frac{k}{R}\,Y \approx k\theta_y.$$

A *finite array of identical holes* in a screen can be represented by an aperture function which is a convolution of a function $f_0(\mathbf{s})$, consisting of delta functions at the positions of the holes, with the aperture function $g(\mathbf{s})$ appropriate to one hole. We now have

$$f(\mathbf{s}, 0) = f_0(\mathbf{s}) \otimes g(\mathbf{s}) \tag{11.81}$$

$$F(\mathbf{q}, 0) = F_0(\mathbf{q})G(\mathbf{q}) \tag{11.82}$$

$$\sigma = \frac{k^2}{(2\pi)^2}\,|F_0(\mathbf{q})|^2\,|G(\mathbf{q})|^2. \tag{11.83}$$

For instance a *row of circular holes* follows from transforms 3.31 and 3.27, and the diffraction pattern will consist of the ring pattern crossed by a set of bands. A *finite lattice of circular holes* follows from transforms 3.33 and 3.27, and the diffraction pattern consists of a ring pattern crossed by a lattice of lines, with bright spots at the lattice intersections.

A *diffracting screen of infinite extent* is treated using eqn (11.39). If the screen possesses a regularity then $F(\mathbf{q}, 0)$ will contain delta functions, and a power spectrum may be derived using the two-dimensional version of eqn (4.27). We obtain the following:

if
$$F(\mathbf{q}, 0) = \sum_n a_n \delta(\mathbf{q} - \mathbf{q}_n) \tag{11.84}$$

then

$$I = \frac{k^2}{(2\pi)^4} \sum_n |a_x|^2 \, \delta(\mathbf{q} - \mathbf{q}_n). \tag{11.85}$$

Equation (11.85) is intended to apply equally to point delta functions and to line delta functions. The integrated intensity in a diffraction spot (or per unit length of a line) will be given by a local integration as follows:

$$\int_{\text{loc}} I \, d\Omega = \int_{\text{loc}} \frac{I}{k^2} \, d\mathbf{q} = \frac{|a_n|^2}{(2\pi)^4} \, . \tag{11.86}$$

As a special case of such regularity we may consider an *infinite, regular, array of identical apertures*. Regarding this as a convolution we may proceed as in eqns (11.81) and (11.82) with $F_0(\mathbf{q})$ a set of delta functions. We now have

$$f(\mathbf{s}, 0) = f_0(\mathbf{s}) \otimes g(\mathbf{s})$$

$$F(\mathbf{q}, 0) = G(\mathbf{q}) \sum_n a_n \delta(\mathbf{q} - \mathbf{q}_n) \tag{11.87}$$

and

$$I = \frac{k^2}{(2\pi)^4} |G(\mathbf{q})|^2 \, \Sigma |a_n|^2 \delta(\mathbf{q} - \mathbf{q}_n). \tag{11.88}$$

An *infinite row of apertures* and an *infinite lattice of apertures* may now be dealt with using eqns (3.35) or (3.36), giving respectively a set of lines and a lattice of spots, the intensities being governed by the function $G(\mathbf{q})$.

If the infinite lattice of apertures just considered is truncated by some function $h(\mathbf{s})$ so that the lattice becomes of finite extent, then the sharply defined directions of diffraction maxima become blurred, as may be seen from the following:

$$f(\mathbf{s}, 0) = [f_0(\mathbf{s}) \otimes g(\mathbf{s})] h(\mathbf{s}) \tag{11.89}$$

$$F(\mathbf{q}, 0) = \frac{1}{(2\pi)^2} [F_0(\mathbf{q})G(\mathbf{q})] \otimes H(\mathbf{q}) \tag{11.90}$$

$$= \frac{1}{(2\pi)^2} \left[\sum_n a_n \delta(\mathbf{q} - \mathbf{q}_n) G(\mathbf{q}_n) \right] \otimes H(\mathbf{q}) \tag{11.91}$$

$$= \frac{1}{(2\pi)^2} \sum_n [a_n G(\mathbf{q}_n) H(\mathbf{q} - \mathbf{q}_n)] \tag{11.92}$$

$$\sigma \approx \frac{k^2}{(2\pi)^4} \sum_n [|a_n|^2 |G(\mathbf{q}_n)|^2 |H(\mathbf{q} - \mathbf{q}_n)|^2]. \tag{11.93}$$

The approximation in eqn (11.93) is valid when the function $H(\mathbf{q} - \mathbf{q}_n)$ is sufficiently sharply peaked not to overlap a neighbouring delta function, which means that the lattice of apertures must be large in size compared with the spacing between apertures.

11.11   The Optical Transfer and Point Spread Functions

A comparison of eqns (11.28) and (11.29) with eqns (7.2) and (7.3) leads naturally to the idea of treating the diffraction problem as a linear system in the sense described in Chapter 7. $f(\mathbf{s}, 0)$ is now the 'input' to the system and $f(\mathbf{S}, z)$ is the 'output'. The 'transfer function' $H(q)$ and the 'impulse response', $h(\mathbf{s})$ are now given by

$$H(\mathbf{q}) = \exp(iz\sqrt{k^2 - q^2}) \tag{11.94}$$

$$h(\mathbf{s}) = \frac{1}{(2\pi)^2} \iint \exp(i\mathbf{q} \cdot \mathbf{s}) \exp(iz\sqrt{k^2 - q^2}) \, d\mathbf{q}. \tag{11.95}$$

We shall adopt the names, common in the optical context, of optical transfer function for $H(\mathbf{q})$ and point spread function for $h(\mathbf{s})$.

It is natural to wonder whether any optical system may be treated as a 'black box' and whether the aperture functions on any two planes may be related using an appropriate optical transfer function or point spread function. As we shall see this is indeed the case if a few simplifying conditions are obeyed. For simplicity let us first consider the image and object planes of a system of unit magnification. Diffractive or other aberrations will perhaps lead to a blurred image, but if the system is linear and shift invariant in the sense described in Chapter 7 then the aperture functions in the image and object planes, $f_{im}(\mathbf{S})$ and $f_{ob}(\mathbf{s})$ will be related by

$$f_{im}(\mathbf{S}) = f_{ob}(\mathbf{S}) \otimes h(\mathbf{S}) \tag{11.96}$$

where $h(\mathbf{S})$ is the result of a delta function amplitude disturbance in the object plane. Note that the shift invariance requirement will be violated if any optical distortions occur which distort rather than blurr the image. We will call an optical system ideal if it is linear and shift invariant even if blurring of the type described by eqn (11.96) occurs. If the magnification is not unity then the shift invariance requirement fails, but a simple process of scaling allows us to modify eqn (11.96) in a straightforward fashion provided that displacement of a point object in the object plane merely displaces the image proportionately. If the magnification is M we now have

$$f_{\text{im}}(\mathbf{S}) = f_{\text{ob}}\left(\frac{\mathbf{S}}{M}\right) \otimes h(\mathbf{S}) \tag{11.97}$$

so that on Fourier transformation we have (see eqn (3.10))

$$F_{\text{im}}(\mathbf{q}) = M^2 F_{\text{ob}}(M\mathbf{q})H(\mathbf{q}). \tag{11.98}$$

The optical system may be thought of as a device for compressing or expanding the angular spectrum of the aperture function (magnification implying compression of the angular spectrum). The diffractive aberrations represented by $H(\mathbf{q})$ achieve their blurring effect merely by modifying the various components in the angular spectrum.

11.12   Diffraction at a Lens

As a preliminary step to calculating the aperture functions in any two planes on either side of a lens it is convenient first to establish that the aperture functions at the two focal planes on either side of a convex lens are related to each other through a Fourier transform. Such a lens may thus be regarded as a device for Fourier transforming a function. The evolution of an aperture function as we consider successive planes in an optical system may be described as follows. As we progress through free space by a distance z the aperture function becomes convoluted with the function $h(\mathbf{s})$ as given in eqn (11.95). On arriving at the focal plane of a lens we imagine the aperture function transferred to the other focal plane and converted to its Fourier transform, perhaps with a change of scale. We then proceed onwards through free space using a convolution as just described until we arrive at the focal plane of another lens, and so on. The normal laws of image formation can be derived using this procedure.

It is well known experimentally that a convex lens focuses parallel incident rays to a point in the focal plane, rays of a particular direction being concentrated into a corresponding point. It is thus plausible that the spatial distribution of amplitude in the focal plane will be related to the angular spectrum of the incident wavefront. Let us now establish this result more rigorously.

Consider a thin double convex spherical lens with radii of curvature r_1 and r_2 having a maximum thickness t_0 and a refractive index n. The thickness of the lens at a distance s from the axis is

$$\left\{ t_0 - s^2 \left(\frac{1}{r_1} + \frac{1}{r_2} \right) \right\}$$

for the case $s \ll r_1$, $s \ll r_2$. The lens thus acts as a device for delaying the phase of a wavefront passing through it by an amount dependent on s^2. If we define the focal length f of the lens by the following relation well known in geometrical optics

$$\frac{1}{2f} = (n - 1)\left(\frac{1}{r_1} + \frac{1}{r_2} \right) \tag{11.99}$$

then it is a simple matter to establish that the aperture function $f_2(\mathbf{s})$ just after the lens is related to that which would have been present in the same plane in the absence of the lens, $f_1(\mathbf{s})$, by the relation

$$f_2(\mathbf{s}) = f_1(\mathbf{s}) \exp \{ ik(n - 1)t_0 - iks^2/2f \}. \tag{11.100}$$

We are assuming that $f_2(\mathbf{s})$ and $f_1(\mathbf{s})$ are zero outside the area covered by the lens. Using the point spread function we are now in a position to find the disturbance in any subsequent plane. The general case is somewhat intractable, but using the Fresnel approximation of eqn (11.32) we may find the disturbance in the focal plane as follows. To simplify the algebra we omit the factor $\exp \{ ik(n - 1)t_0 \}$ which is unimportant for thin lenses. If $f_3(\mathbf{s})$ is the disturbance in the focal plane we have, using eqns (11.32) and (11.100):

$$f_3(\mathbf{S}) = -\frac{ik}{2\pi f} \exp \left[ik(f + S^2/2f) \right] \iint \exp \left(-ik\mathbf{S} \cdot \mathbf{s}/f \right)$$

$$\times \{ f_2(\mathbf{s}) \exp (iks^2/2f) \} \, d\mathbf{s} \tag{11.101}$$

$$= -\frac{ik}{2\pi f} \exp \left[ik(f + S^2/2f) \right] \iint \exp \left(-ik\mathbf{S} \cdot \mathbf{s}/f \right) f_1(\mathbf{s}) \, d\mathbf{s} \tag{11.102}$$

$$= -\frac{ik}{2\pi f} \exp \left[ik(f + S^2/2f) \right] F_1\left(\frac{k}{f} \mathbf{S} \right). \tag{11.103}$$

This shows that $f_3(\mathbf{S})$ is a phase distorted and scaled version of the Fourier transform of $f_1(\mathbf{s})$. We now use eqn (11.30) to work backwards and relate the angular spectrum $F_1(\mathbf{q})$ to that in the preceding first focal plane of the lens $F_0(\mathbf{q})$ giving:

$$F_1(\mathbf{q}) = \exp\left[i(kf - fq^2/2k)\right] F_0(\mathbf{q}).\qquad(11.104)$$

On substituting this result into eqn (11.103) we obtain, with $\mathbf{q} = k\mathbf{S}/f$,

$$f_3(\mathbf{S}) = -\frac{ik}{2\pi f}\, e^{2ikf} F_0\!\left(\frac{k}{f}\,\mathbf{S}\right).\qquad(11.105)$$

Equation (11.105) confirms the important result that apart from constant factors and from a change of scale, the aperture functions in the two focal planes form a Fourier pair.

The ordinary rules of geometrical optics follow as special cases using this technique. For instance the disturbances in two planes which are respectively distances z_1 before the first focal plane and z_2 beyond the second focal plane come to be equal if $z_1 z_2 = f^2$ (apart from inversion, scaling and constant phase factors). This corresponds to image formation, the relation $z_1 z_2 = f^2$ being familiar in geometrical optics.

The formation of an image by two similar lenses may clearly be described as follows. If an object is put at the focal plane of one lens a Fourier image is formed at the other focal plane. If this second plane is now made the first focal plane of the other lens then clearly by Fourier transforming a second time an inverted image of the original object is recovered at the second focal plane of the second lens, since as shown in Appendix D two successive Fourier transforms merely invert a function. This system is in fact very similar to that employed in a microscope (or with a magnifying glass if the lens of the eye is considered to be the second lens). The effect of a stop at the central focal plane in such a system is especially easy to determine. The stop modifies the transform of the function, so that the reconstructed image is convoluted with the transform of the transmission function of the stop. This method of calculating the diffractive aberration due to a stop was introduced by Abbe. As an example consider a circular hole of radius R in a stop at the central focal plane of such a system. The transmission function $U(\mathbf{s})$ of the stop is

$$U(\mathbf{s}) = 1\qquad [|\mathbf{s}| < R]$$

$$= 0\qquad [|\mathbf{s}| > R]\qquad(11.106)$$

and the point spread function applicable to the final focal plane is given (apart from constant factors) by

$$h(S) \propto \iint \exp\left(-ikS \cdot s/f\right) U(s)\, ds \qquad (11.107)$$

$$\propto 2\pi R^2 \frac{J_1(kRS/f)}{(kRS/f)}. \qquad (11.108)$$

By using other forms of stop in which the phase or amplitude is modified in other ways than that of eqn (11.106) one may of course produce spread functions $h(S)$ which differ from the form given in eqn (11.108), even if the restriction that $U(s) = 0$ for $s > R$ is adhered to. Such a process is called apodization and improvements in the resolving power of such a system can be produced by this means in that the radius of the first zero in amplitude of $h(S)$ can be reduced. One is not of course violating the uncertainty principle which relates the mean square widths of $U(s)$ and $h(S)$.

The two lens system just discussed forms the basis of the phase contrast microscope invented by Zernike. With this device objects which are transparent and which affect the phase slightly but not the amplitude of the illuminating wavefront (and whose image is thus normally invisible) are rendered visible in the image plane by the insertion of a subtly devised stop at the central focal plane. The result is to convert the weak phase modulation to a proportionate amplitude modulation. To expose the principle of the method let the two lenses have equal focal lengths, and let us omit constant phase factors. Let $f_0(s)$, $f_1(s)$, $f_2(s)$ and $f_3(s)$ be the respective wavefront disturbances in the object plane, just before the central stop, just after it, and in the image plane. Let $f_0(s)$ be of the form $\exp[-ig(s)]$ where the function $g(s)$ represents the phase modulation produced by the object. We assume $g(s) \ll 1$. In the absence of a stop we have image formation as follows:

$$f_0(s) \propto \exp[-ig(s)] \propto \{1 - ig(s)\} \qquad (11.109)$$

$$f_1(s) = f_2(s) \propto \left\{2\pi\delta\left(\frac{k}{f}s\right) - iG\left(\frac{k}{f}s\right)\right\} \qquad (11.110)$$

$$f_3(s) \propto \{1 - ig(-s)\} \propto \exp[-ig(-s)]. \qquad (11.111)$$

In the central focal plane there is thus a delta function at $s = 0$, with a distribution round it governed by the Fourier transform of $g(s)$. If we now insert a stop which covers the central delta function only, which delays it in phase by $3\pi/2$ (i.e. advances it by $\pi/2$) and reduces its amplitude by a factor A, then the expressions now become:

$$f_0(\mathbf{s}) \propto \exp\left[-ig(\mathbf{s})\right] \propto \{1 - ig(\mathbf{s})\} \tag{11.112}$$

$$f_1(\mathbf{s}) \propto \left\{2\pi\delta\left(\frac{k}{f}\mathbf{s}\right) - iG\left(\frac{k}{f}\mathbf{s}\right)\right\} \tag{11.113}$$

$$f_2(\mathbf{s}) \propto \left\{2\pi iA\delta\left(\frac{k}{f}\mathbf{s}\right) - iG\left(\frac{k}{f}\mathbf{s}\right)\right\} \tag{11.114}$$

$$f_3(\mathbf{s}) \propto i\{A - g(-\mathbf{s})\}. \tag{11.115}$$

Clearly the intensity in the final image is proportional to $\{A - g(-\mathbf{s})\}^2$ which corresponds to a visible image. The simple procedure of a totally absorbing stop, with $A = 0$, gives a 'negative' image the phase object appearing bright on a dark background. If conversely the stop alters the phase appropriately but produces no absorption of the delta function then we have $A = 1$ and a 'positive' image is formed but with low contrast (since $g(\mathbf{s}) \ll 1$). If A is reduced to the smallest possible value, consistent with $\{A - g(\mathbf{s})\}$ being everywhere positive, then one has a 'positive' image with maximum contrast. It should be noted that the delta function is in practice broadened due to the finite size of the lenses so that a finite size of stop is required.

Chapter 12

Optical Coherence and Holography

12.1 Introduction

The optical theory described in Chapter 11 has assumed throughout a coherent radiation field (see Section 11.3, eqn (11.3)). In practical terms this is a severe limitation and represents a condition never attained completely. We start this chapter by considering the meaning and characterisation of coherence, and of the effects of not having perfect coherence. This discussion is exemplified by a discussion of the famous experiments conducted by Michelson and by Hanbury Brown and Twiss. We follow this by a discussion of various modern optical techniques associated with holography. Our discussion throughout is aimed at showing how Fourier transform techniques expose the basic principles involved in a useful way. For a more rigorous approach the reader is referred to reference 1, whilst for more experimental details on holography the reader is referred, for instance, to references 14 and 19.

Let us start with a qualitative discussion of coherence. A coherent wave field is a monochromatic one in which although the phase and amplitude of the radiation may perhaps vary from point to point, neither of these quantities varies with time. If on the other hand the phase and perhaps the amplitude at any point is subject to fluctuations which are a function of time then the ideal coherence is lost. One result of course, as the discussion of Sections 6.6 and 6.7 shows, is that the radiation is no longer strictly monochromatic. If however the phase or amplitude fluctuations occur slowly compared with the fundamental harmonic oscillations then the corresponding spread of frequencies is small compared with the basic frequency and we shall describe such a wave field as being *quasi-mono-chromatic*. Such random fluctuations in optical wave fields are the rule rather than the exception when using incandescent or discharge tube sources. An idealized model of such sources is to consider them as emitting a series of wave trains of finite duration at random instants of time, just as in shot noise (see Section 6.5). Doppler effects can mean that in addition to the frequency spread due to the finite duration of each wave train there is a further frequency spread due to thermal motions.

Fluctuations such as these can of course have a profound effect on diffraction phenomena, in that if an aperture function $f(s, 0)$ like that introduced in Section 11.3 is fluctuating in time then the disturbance $f(S, z)$ in some subsequent plane will also be fluctuating with time. Even if the fluctuations in the $z = 0$ plane are fluctuations in phase only this can nevertheless have drastic effects on the intensity distribution in, say, a Fraunhofer pattern so that the shape of the resulting intensity diffraction pattern may be fluctuating 'wildly' with time. Detectors such as the eye and the photographic plate respond relatively slowly to changes of intensity and they thus measure only a time averaged intensity pattern.

We shall restrict discussion to the fluctuations associated with a quasi-monochromatic wave field and we may note that this means that all the equations of Chapter 11 remain valid as representations of what is happening 'instantaneously' over a period not longer than about ten or a hundred periods of oscillation. This means that we may readily generalize the equations derived in Chapter 11 by writing the complex amplitudes as explicit functions of time, say as $f(s, 0, t)$ and as $f(S, z, t)$ in the planes at the aperture and at z respectively. If the fluctuations over the aperture plane all occur synchronously together so that $f(s, 0, t)$ may be split up and written as $f(s, 0)f_1(t)$ then the disturbance in a subsequent plane will also be separable as $f(S, z)f_1(t)$ where the functions $f(S, z)$ and $f(s, 0)$ are related just as in Chapter 11. The spatial distribution of amplitude and intensity in the Fresnel or Fraunhofer field is thus unaltered by such fluctuations and the radiation field is said to be *laterally coherent* across the aperture plane. If on the other hand the fluctuations at two points in the aperture plane are large and completely uncorrelated (for points more than about a wavelength apart) then the wave front is said to be *laterally incoherent*. Clearly situations in between these extremes can occur and a lateral coherence length can be defined which represents the approximate distance over which local fluctuations are correlated, and beyond which they become uncorrelated. Diffraction phenomena are fully developed if the aperture function is laterally coherent, and they become blurred and finally cease to exist as the wave front is progressively made laterally incoherent. A wave front which is laterally coherent over useful macroscopic distances (i.e. several mm) may be obtained directly from a laser, or from an incoherent wave front by limiting it at an aperture and allowing the wave front to progress a suitable distance beyond the aperture. This latter method is the classical one associated with the name Young whereby a pin-hole is used as the primary source of light.

Independently of whether a wave front is laterally coherent or not one can define analogously a *coherence time* as, roughly speaking, the period over which the complex amplitude remains constant but beyond which will have changed in a random manner. The spectral width of the radiation in terms of frequency is approximately equal to the inverse of this coherence time, as is clear from the discussion in Sections 2.4(f) and in 6.6 and 6.7. Associated with the coherence time is a *longitudinal coherence length* which is the corresponding distance along the direction of propagation within which the random fluctuations are correlated. Clearly the longitudinal coherence length is simply equal to the product of the speed of propagation and the coherence time. The coherence time and coherence length are important in those interference experiments in which a wave front is divided and a delayed portion is made to 'interfere' with an undelayed portion. As the delay is increased beyond the coherence time so the interference phenomena become less pronounced and eventually cease. A high pressure mercury lamp gives a coherence length of about 0.1 mm, whilst lasers can give longitudinal coherence lengths of several kilometres.

12.2   The Characterisation of Coherence

We are concerned with the correlation in time of the disturbances at different parts of the wave field, and a cross-correlation function of the type introduced in Section 5.2 is a natural way of doing this. In fact for two points at S_1 and S_2 on a plane at z (using the co-ordinate system described in Section 11.2) we define the *complex degree of coherence* $\Gamma_{12}(t)$ as:

$$\Gamma_{12}(t) = \frac{\langle f^*(S_1, t_1) f(S_2, t_1 + t) \rangle}{f_1 f_2} \qquad (12.1)$$

where the triangular brackets refer to averages over t_1 and where f_1 and f_2 are real quantities defined by:

$$\begin{aligned} f_1{}^2 &= \langle |f(S_1, t_1)|^2 \rangle \\ f_2{}^2 &= \langle |f(S_2, t_1)|^2 \rangle . \end{aligned} \qquad (12.2)$$

The quantities f_1 and f_2 are introduced for normalization purposes so that $\Gamma_{12}(t)$ has a modulus varying between zero for complete incoherence to unity for complete coherence. $\Gamma_{12}(t)$ is sometimes written in terms of real quantities $\gamma_{12}(t)$ and θ as follows:

$$\Gamma_{12}(t) = \gamma_{12}(t)\,e^{i\theta}. \tag{12.3}$$

The real quantity $\gamma_{12}(t)$ is known as the *degree of coherence* between the two points, and θ represents the effect of an average phase difference between the two points. It is not immediately obvious that $\gamma_{12}(t)$ has a maximum value of unity, but this can be verified using the Schwartz inequality (see eqns F.1 and F.6).

Sometimes we are concerned mainly with lateral coherence on a plane perpendicular to the direction of propagation, and then the quantity $\Gamma_{12}(0)$ is of interest, which we may write as Γ_{12} or as $\Gamma(\mathbf{S}_1, \mathbf{S}_2)$ to stress its dependence on $\mathbf{S}_1$ and $\mathbf{S}_2$. Analogously if we are concerned primarily with time coherence the quantity of interest is $\Gamma_{11}(t)$ written simply as $\Gamma(t)$. Clearly

$$\Gamma(\mathbf{S}_1, \mathbf{S}_2) \to 1 \quad \text{as} \quad (\mathbf{S}_1 - \mathbf{S}_2) \to 0 \tag{12.4}$$

$$\Gamma(t) \to 1 \quad \text{as} \quad t \to 0. \tag{12.5}$$

A lateral coherence length or a coherence time can be defined in terms of some appropriate width for each of the functions $\gamma(\mathbf{S}_1, \mathbf{S}_2)$ or $\gamma(t)$.

Note that although we have made the reasonable tacit assumption that the time variation is a stationary random process, this has not been assumed to hold with respect to spatial variations. Thus we allow for the possibility that $\Gamma_{12}(t)$ may depend on the values of $\mathbf{S}_1$ and $\mathbf{S}_2$ and not merely on their difference $(\mathbf{S}_1 - \mathbf{S}_2)$.

12.3 Visibility of Interference Fringes

The quantity $\gamma_{12}(0)$ describing the degree of coherence between two points has a direct experimental interpretation in terms of the visibility of the interference fringes in a two hole interference experiment. Let us consider the time average Fraunhofer diffraction intensity pattern in an experiment with two holes at $\mathbf{s}_1$ and $\mathbf{s}_2$ in a screen in the $z = 0$ plane illuminated by a wave front which is only partially coherent relative to $\mathbf{s}_1$ and $\mathbf{s}_2$. We assume the holes to be small in size compared with their separation and small compared with the lateral coherence length. Let the holes each have an area A.

From eqn (11.34) the disturbance in a plane at z is:

$$f(\mathbf{S}, z, t) = -\frac{ik}{2\pi z}\,e^{ikR} \iint \exp\left(-i\frac{k}{z}\,\mathbf{s}\cdot\mathbf{S}\right) f(\mathbf{s}, 0, t)\,d\mathbf{s} \tag{12.6}$$

$$= -\frac{ik}{2\pi z}Ae^{ikR}\left\{f(\mathbf{s}_1, t)\exp\left(-i\frac{k}{z}\mathbf{s}_1 \cdot \mathbf{S}\right)\right.$$

$$\left. + f(\mathbf{s}_2, t)\exp\left(-i\frac{k}{z}\mathbf{s}_2 \cdot \mathbf{S}\right)\right\}. \tag{12.7}$$

The time averaged intensity in the z plane is thus (omitting a constant proportionality term)

$$\langle|f(\mathbf{S}, z, t)|^2\rangle \propto \langle|f(\mathbf{s}_1, t)|^2\rangle + \langle|f(\mathbf{s}_2, t)|^2\rangle$$

$$+ \langle f(\mathbf{s}_1, t)f^*(\mathbf{s}_2, t)\rangle \exp\left[-i\frac{k}{z}(\mathbf{s}_1 - \mathbf{s}_2)\cdot\mathbf{S}\right]$$

$$+ \langle f^*(\mathbf{s}_1, t)f(\mathbf{s}_2, t)\rangle \exp\left[+i\frac{k}{z}(\mathbf{s}_1 - \mathbf{s}_2)\cdot\mathbf{S}\right] \tag{12.8}$$

$$\propto f_1{}^2 + f_2{}^2 + 2f_1 f_2 \gamma_{12} \cos\left\{\frac{k}{z}(\mathbf{s}_1 - \mathbf{s}_2)\cdot\mathbf{S} + \theta\right\}. \tag{12.9}$$

We have introduced quantities f_1 and f_2 as in eqn (12.2). We see that the time averaged intensity pattern consists of a uniform background with a striped pattern superposed, the stripes being perpendicular to $(\mathbf{s}_1 - \mathbf{s}_2)$. The term θ merely displaces the pattern as expected. The visibility of the fringes clearly depends on the 'contrast' in the fringes and is often defined in terms of maximum and minimum intensities $I_{\max}$ and $I_{\min}$ as below:

$$\text{Visibility} = \frac{I_{\max} - I_{\min}}{I_{\max} + I_{\min}}. \tag{12.10}$$

From eqn (12.9) we see that the visibility is given by

$$\text{Visibility} = \frac{2f_1 f_2}{f_1{}^2 + f_2{}^2}\gamma_{12}. \tag{12.11}$$

Thus the degree of coherence has a very direct interpretation in such an experiment. Indeed a series of experiments with holes in various positions allows one in principle to plot out the complete function $\gamma(\mathbf{s}_1, \mathbf{s}_2)$. This was the basis of Michelson's famous stellar interferometer with which he investigated the lateral coherence of the radiation arriving from a star or pair of stars. From a knowledge of γ_{12} the angular diameter of the star could be calculated according to the ideas to be discussed in Section 12.6.

A system of mirrors allowed effective aperture separations of several metres, without using correspondingly large lenses, in forming the Fraunhofer pattern.

Analogously in the ordinary Michelson interferometer in which a wave front is split and the two portions superposed after introducing a relative delay t, one may measure $\gamma(t)$ by measuring the visibility of the interference phenomena using an approach very similar to that just given.

12.4   Effect of Partially Coherent Illumination on Diffraction Patterns

The two hole aperture considered in the previous section is of course a very special case. We consider now the effect of illuminating an aperture of any sort with partially coherent radiation, with the simplifying assumption that the degree of coherence $\Gamma(s_1, s_2)$ in the incident radiation depends on $s(= s_1 - s_2)$ and not on s_1. We will accordingly write $\Gamma(s_1, s_2)$ as $\Gamma(s)$, and our aim will be to show that the resulting intensity diffraction pattern is simply given by a convolution between the pattern obtained in coherent illumination with a function related to the spatial Fourier transform of $\Gamma(s)$.

Consider the disturbance $f(s, 0, t)$ emerging from an aperture when a partially coherent incident wave front $h(s, t)$ of unit amplitude is incident on a screen having transmission function $t(s)$. Thus

$$f(s, 0, t) = h(s, t)t(s). \tag{12.12}$$

The Fraunhofer intensity distribution depends on the spatial energy spectrum $S_f(q, t)$ of $f(s, 0, t)$ where q is a variable conjugate to s (eqn (11.38)), and we will now proceed to find the time average of $S_f(q, t)$ by first finding the spatial autocorrelation function $\rho_f(s, t)$ of $f(s, 0, t)$ and Fourier transforming this to give $S_f(q, t)$ using the Wiener–Khintchine theorem (eqn (5.31)). If we use triangular brackets to represent a time average then

$$\langle \rho_f(s, t) \rangle = \left\langle \iint h^*(s_1, t)t^*(s_1)h(s_1 + s, t)t(s_1 + s)\, ds_1 \right\rangle \tag{12.13}$$

$$= \iint \Gamma(s)t^*(s_1)t(s_1 + s)\, ds_1 \tag{12.14}$$

$$= \Gamma(s)\rho_t(s) \tag{12.15}$$

where $\rho_t(s)$ is the autocorrelation function of $t(s)$. If now we write $G(q)$ as the Fourier transform of $\Gamma(s)$, and $S_t(q)$ as the Fourier transform of $\rho_t(s)$

we get on Fourier transforming eqn (12.15)

$$(2\pi)^2 \langle S_f(\mathbf{q}, t) \rangle = G(\mathbf{q}) \otimes S_t(\mathbf{q}). \tag{12.16}$$

We may now finally use eqn (11.38) to give the angular distribution of intensity in the Fraunhofer pattern as

$$\sigma(\mathbf{q}) = \frac{k^2}{(2\pi)^4} G(\mathbf{q}) \otimes S_t(\mathbf{q}). \tag{12.17}$$

Equation (12.17) illustrates the result we wished to prove since $S_t(\mathbf{q})$ describes the pattern which would occur with laterally coherent incident radiation, and the convolution with $G(\mathbf{q})$ represents a blurring.

12.5   The Hanbury Brown and Twiss Experiment

This ingenious technique provides an alternative to the Michelson stellar interferometer for investigating the lateral coherence of the wave front arriving from a star. In order to investigate a wave front $f(\mathbf{s}, t)$ two photo-multipliers are placed at positions $\mathbf{s}_1$ and $\mathbf{s}_2$ which give output currents $I_1(t)$ and $I_2(t)$ proportional (ideally) to the intensities at $\mathbf{s}_1$ and $\mathbf{s}_2$ so that

$$I_1(t) \propto |f(\mathbf{s}_1, t)|^2,$$
$$I_2(t) \propto |f(\mathbf{s}_2, t)|^2. \tag{12.18}$$

These fluctuating currents are multiplied together and the time averaged product measured as a function of $\mathbf{s}_1$ and $\mathbf{s}_2$. Clearly for $(\mathbf{s}_1 - \mathbf{s}_2)$ less than the lateral coherence length the fluctuations are synchronous so that if the general intensity is uniform then

$$\langle I_1(t)I_2(t) \rangle = \langle I_1^2(t) \rangle = \langle I_2^2(t) \rangle. \tag{12.19}$$

On the other hand for separations large compared with the coherence length the fluctuations are independent and

$$\langle I_1(t)I_2(t) \rangle = \langle I_1(t) \rangle^2 = \langle I_2(t) \rangle^2. \tag{12.20}$$

Clearly if in addition one can introduce a delay between the two currents before forming the product one may measure a coherence function $\Omega_{12}(t)$ which describes intensity correlations just as $\gamma_{12}(t)$ describes correlations in complex amplitude. The lateral coherence lengths associated with $\Omega_{12}(0)$ and $\gamma_{12}(0)$ will naturally be roughly the same, and in fact if certain assumptions are made about the statistical nature of the fluctuations Wolf has shown that $\Omega_{12}(t)$ is proportional to the square of $\gamma_{12}(t)$.

The theory given here is a purely classical one, and at low intensities the quantum nature of the phenomenon needs to be considered. Such a discussion is however outside our scope.

There is a close analogy between the process carried out in the Hanbury Brown and Twiss experiment and another class of experiments in which a polychromatic light beam is detected by a photomultiplier and the resulting current is analysed to give either a power spectrum or a time autocorrelation function. The photomultiplier is a non-linear device (in the sense used in Chapter 7) and a detailed analysis is outside our scope. However we will quote one useful and important result. Suppose we are dealing with a real stationary random function $f(t)$, which may for instance represent the magnitude of the electric vector in a light beam. Let $I(t)$ be the resulting intensity, proportional to $f^2(t)$, and let $\rho_f(t), \rho_I(t), P_f(\omega)$ and $P_I(\omega)$ be the various autocorrelation functions and power spectra. With certain assumptions about the statistical nature of the fluctuations it is shown in reference 18, for instance, that

$$P_I(\omega) \propto P_f(\omega) \otimes P_f(\omega)$$

so that

$$\rho_I(t) \propto \{\rho_f(t)\}^2.$$

This result may be illustrated by a simple example in which 'beats' occur. Suppose that $f(t)$ is formed by 'beating' together two laser beams at slightly different frequencies ω_1 and ω_2 so that $P_f(\omega)$ contains peaks of narrow but finite width at $\pm \omega_1$, and at $\pm \omega_2$. Some thought soon establishes that $P_f(\omega) \otimes P_f(\omega)$ will contain broadened peaks centred on the frequencies $\pm (\omega_2 - \omega_1), \pm 2\omega_1, \pm (\omega_1 + \omega_2)$ and $\pm 2\omega_2$ as well as a strong peak centred on zero frequency. The peaks at $\pm(\omega_2 - \omega_1)$ correspond simply to beats between the various frequency components of one beam with those in the other beam. The peak centred on zero frequency corresponds to beats between the various components within one and the same laser beam, whilst the other high frequency terms correspond to the second harmonic which quadratic detectors always produce. This analysis has shown that although perfectly regular beats may not exist between two incoherent sources, nevertheless the techniques of autocorrelation and spectral analysis can be used to reveal equivalent information.

12.6 The Propagation of Coherence

The lateral coherence characteristics of a wave front alter as it progresses and, for instance, a disturbance which is incoherent in the $z = 0$ plane develops a larger and larger lateral coherence length as larger values of z are considered. We will now investigate this and show that the lateral coherence properties in a plane at z in the radiation field produced by incoherent radiation emerging from an aperture in the $z = 0$ plane are closely related to the diffraction pattern that would be produced by coherent radiation emerging from the aperture.

Let the complex amplitudes in the plane at z and in the $z = 0$ plane be $f(S, z, t)$ and $f(s, t)$ respectively. Let us seek an expression for the complex degree of coherence between a point at ϕ in the z plane, and a point on the axis at $\phi = 0$ in the same plane. We may use eqn (11.32) to give, in the Fresnel approximation, using triangular brackets to represent time averages:

$$\langle f^*(0, z, t)f(S, z, t)\rangle$$

$$= \left(\frac{k}{2\pi z}\right)^2 \exp\left(ikS^2/2z\right) \left\langle \iint f^*(s, t) \exp\left(-iks^2/2z\right) ds \right.$$

$$\times \iint \exp\left(-\frac{ik}{z}\mathbf{S}\cdot\mathbf{s}\right) f(s, t) \exp\left(\frac{iks^2}{2z}\right) ds \right\rangle \qquad (12.21)$$

$$= \left(\frac{k}{2\pi z}\right)^2 \exp\left(ikS^2/2z\right) \left\langle \iiiint \exp\left(-\frac{ik}{z}\mathbf{S}\cdot\mathbf{s}_2\right) f^*(\mathbf{s}_1, t)f(\mathbf{s}_2, t) \right.$$

$$\times \exp\left[ik(s_2{}^2 - s_1{}^2)/2z\right] d\mathbf{s}_1\, d\mathbf{s}_2 \right\rangle \qquad (12.22)$$

$$= \left(\frac{k}{2\pi z}\right)^2 \exp\left(ikS^2/2z\right) \iiiint \exp\left(-\frac{ik}{z}\mathbf{S}\cdot\mathbf{s}_2\right) f_1 f_2 \Gamma(\mathbf{s}_1, \mathbf{s}_2)$$

$$\times \exp\left[ik(s_2{}^2 - s_1{}^2)/2z\right] d\mathbf{s}_1\, d\mathbf{s}_2 \qquad (12.23)$$

where $f_1{}^2$ and $f_2{}^2$ represent the intensities at $\mathbf{s}_1$ and $\mathbf{s}_2$ as in eqn (12.2).

We see already that the lateral coherence characteristics in the z plane and the $z = 0$ plane are related via a Fourier transform. If the radiation in the $z = 0$ plane has a very short coherence length a simplification is possible. Suppose by way of illustration that $\Gamma(\mathbf{s}_1, \mathbf{s}_2)$ has a correlation length L. If L is sufficiently short then integration of eqn (12.23) over $d\mathbf{s}_1$ leaves the following, where A is a small area of order L^2:

$$\langle f^*(0, z, t)f(S, z, t)\rangle = \left(\frac{k}{2\pi z}\right)^2 A \exp\left(ikS^2/2z\right)$$

$$\times \iint \exp\left[-i\frac{k}{z}\mathbf{S}\cdot\mathbf{s}_2\right] f_2{}^2\, d\mathbf{s}_2. \qquad (12.24)$$

Clearly the intensity at $S = 0$ is given from eqn (12.24) as

$$\langle f^*(0, z, t) f(0, z, t) \rangle = \left(\frac{k}{2\pi z}\right)^2 A \iint f_2^2 \, ds_2$$

so that finally the complex degree of lateral coherence in the z plane is given (for $kS^2/2z \ll 1$) by:

$$\Gamma(0, S) = \frac{\iint \exp\left(-i\frac{k}{z} S \cdot s\right) I(s) \, ds}{\iint I(s) \, ds} \qquad (12.25)$$

where we have written the intensity at s in the z plane as $I(s)$. The result is clearly insensitive to the exact behaviour of $\Gamma(s_1, s_2)$ provided it has a short coherence length, since the area A has cancelled out. A closer examination of the required 'shortness' of L shows that we have required

$$\frac{s_m L}{z\lambda} \ll 1 \qquad (12.26)$$

where s_m represents a maximum value of $|s|$ beyond which $f(s, t)$ is essentially zero. The condition in eqn (12.26) ensures that the term involving $(s_2^2 - s_1^2)/2z$ in eqn (12.22) may be ignored.

We see from eqn (12.25) that $\Gamma(0, S)$ varies with S in exactly the same way as would the amplitude in a coherent diffraction experiment with $f(s)$ in the $z = 0$ plane proportional to $I(s)$. This means that for an incoherent source (e.g. a star or a pin-hole) of dimension D then at a distance z away the lateral coherence length is of order $\lambda z/D$. For a uniform circular source of diameter D we find from eqn (11.79) that $\Gamma(0, S)$ first falls to zero at $S = 1 \cdot 22\lambda z/D$.

The Fourier transform in eqn (12.25) provides the basis for measuring stellar dimensions in the stellar interferometers of Michelson and of Hanbury Brown and Twiss. The result expressed is often known as the van Cittert–Zernike theorem after those who originally derived it (using a different approach from that given here). More general treatments of the propagation of coherence exist including the effects of lens systems, and we refer the reader to reference 1.

12.7 Holography—Basic Principle

Holography may be thought of as a method of recording by photographic means the complicated wave front at some arbitrary plane in an optical system and then subsequently reconstructing the wave front from the information stored on the photograph. As a wave front propagates through an optical system a complete characterisation of the wave field is possible on the basis of a knowledge of the complex amplitude $f(S, z)$ at a plane at z. However much of the information is buried crucially in the phase and a photographic plate which responds only to intensity will record only very incomplete information if it is simply placed in the plane under consideration and exposed. Holography is based on an ingenious method of allowing information on amplitude and phase to be recorded using a detecting device which responds only to intensity. The method depends crucially on the coherence properties of the radiation, and for the moment we shall assume completely coherent radiation.

Before showing a typical experimental arrangement let us give a mathematical illustration to expose the principle of the method. If $f(S)$ is the complex amplitude to be recorded (we have dropped the variable z since this will be constant throughout) then the device we adopt is to add to $f(S)$ a wave front with complex amplitude $\exp(i\mathbf{q}_0 \cdot \mathbf{S})$ where $\mathbf{q}_0$ is constant. Such a wave front simply represents a plane wave progressing in a direction which makes an angle θ with the z axis where

$$\sin \theta = q_0/k. \tag{12.27}$$

The resulting amplitude $f_1(S)$ is

$$f_1(S) = f(S) + \exp(i\mathbf{q}_0 \cdot \mathbf{S}) \tag{12.28}$$

so that the intensity is given by

$$f(S)f^*(S) = |f(S)|^2 + 1 + f(S)\exp(-i\mathbf{q}_0 \cdot \mathbf{S})$$
$$+ f^*(S)\exp(i\mathbf{q}_0 \cdot \mathbf{S}). \tag{12.29}$$

If this intensity is recorded on a photographic plate which (rather ideally) produces a transmission function proportional to the intensity, and if this transparency is now illuminated by a plane wave front the following wave front $f_2(S)$ is of course produced:

$$f_2(S) \propto |f(S)|^2 + 1 + f(S)\exp(-i\mathbf{q}_0 \cdot \mathbf{S})$$
$$+ f^*(S)\exp(i\mathbf{q}_0 \cdot \mathbf{S}). \tag{12.30}$$

The third term on the right-hand side of eqn (12.30) is exactly what would be produced if the original wave front had passed through a prism, since the term $\exp(-i\mathbf{q}_0 \cdot \mathbf{S})$ simply represents a progressive retardation of phase across the plane under consideration. This wave front is correct as regards amplitude and phase and a viewer receiving it will 'see' the original scene or object which gave rise to the wave front $f(S)$. The effect is three-dimensional in that different aspects of the scene will be visible as the viewer moves his head. The first two terms on the right-hand side of eqn (12.30) represent a more or less uniform wave front proceeding straight forward, whilst the last term gives rise to what is known as a conjugate image. This conjugate wave front reconstructs $f^*(S)$ instead of $f(S)$, again deflected as though by a prism but into a direction on the opposite side of the z axis. Now the evolution of a wave front $f^*(S)$ as it progresses forwards along the z axis is simply the complex conjugate of what we would obtain if we traced the history of a wave front $f(S)$ backwards along the z axis (as is clear from eqn (11.28) for instance). Thus as it progresses the wave front $f^*(S)$ forms a real (as opposed to virtual) image of the scene from which it emanated. Fortunately due to the prism like deflection effects the ordinary and conjugate wave fronts become separated and the required aim of reconstructing $f(S)$ has been achieved. As we will show in a moment the process is still realisable even if the photographic process departs from the ideal assumed above.

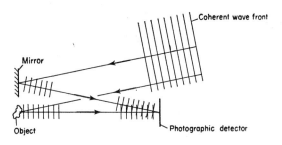

FIG. 12.1. Production of a hologram.

The essence of one method of producing a hologram of the wave front reflected from some object is shown in Fig. 12.1. The reference wave front, $\exp(iq_0 \cdot S)$, is produced by reflecting part of a coherent illuminating wave front from a suitably oriented mirror. The other part of the wave front is reflected off the object of interest, and the hologram is produced in a region where the two wave fronts superpose.

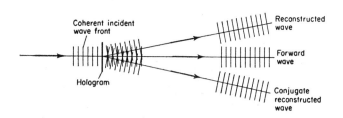

FIG. 12.2. Reconstruction of wave front from a hologram.

The reconstruction of the wave front from the resulting hologram may be achieved by illuminating the hologram with a plane coherent wave front as in Fig. 12.2.

To show rather more vividly the difference between the ordinary and conjugate wave fronts we show in Fig. 12.3 what the two sets of wave fronts would be like if the original object (in Fig. 12.1) had been simply a point object radiating spherical wave fronts. We see in Fig. 12.3 that the ordinary reconstructed wave front diverges as from a virtual point source whilst the conjugate wave front converges to give a real point image.

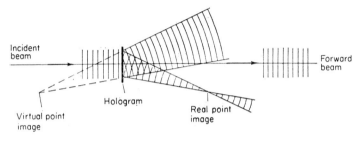

FIG. 12.3. Reconstruction of wave fronts using a point object.

We have so far ignored the fact that in reality the transmission of a photographic transparency is not proportional to the intensity of the illumination during the original exposure. Let us now consider this. In fact over a certain range of intensities the intensity transmittance T_I of a photographic transparency is related to the integrated intensity of exposure I by a relation of the form

$$T_I \propto I^{-\gamma} \qquad (12.31)$$

where γ is a constant. The amplitude transmittance is thus governed by $\gamma/2$ and we obtain from eqn (12.29) that the transmittance $T(S)$ of the hologram will be

$$T(S) \propto \{1 + |f(S)|^2 + f(S)\exp(-i\mathbf{q_0} \cdot S) + f^*(S)\exp(i\mathbf{q_0} \cdot S)\}^{-\gamma/2}. \quad (12.32)$$

If we arrange that the reference beam is intense as compared with the wave front $f(S)$ then we may use part of a binomial expansion to approximate eqn (12.32) as

$$T(S) \propto \left\{1 - \frac{\gamma}{2}|f(S)|^2 - \frac{\gamma}{2}f(S)\exp(-i\mathbf{q_0} \cdot S)\right.$$

$$\left. - \frac{\gamma}{2}f^*(S)\exp(i\mathbf{q_0} \cdot S)\right\} \qquad (12.33)$$

and we see that in this approximation the efficacy of the method is unimpaired.

12.8   Fourier versus Fresnel Holography

Various ingenious tricks are possible using the holographic technique. For instance different objects can be placed in the apparatus in turn and a composite hologram formed by superposing all the exposures on to one photographic plate. On reconstruction the objects will appear superposed with the resultant amplitudes (not intensities) added, so that interference fringes can in principle be formed between the different objects. This can be useful in showing up small mechanical movements or strains: a composite hologram from a strained and unstrained object and the resulting holographic image is crossed by interference fringes in the region of strain. A diminished image results if during reconstruction one uses divergent radiation from a point source. In this case if the point source is a distance z from the hologram then the illuminating wave front at the hologram is of the form $\exp(ikS^2/2z)$. With this illumination we obtain, from the third term on the right-hand side of eqn (12.33), a reconstructed wave of the form

$$f(S)\exp(-i\mathbf{q}_0\mathbf{S})\exp(ikS^2/2z). \tag{12.34}$$

The result is just what would be obtained if the wave front not only passed through a prism but also passed through a concave lens (see eqn (11.100)). The conjugate image can thus be magnified, whilst the virtual image is diminished.

It is interesting now to consider what will happen if we use a point source of radiation for the reference beam during the formation of the hologram. The reference waveform to be inserted into eqn (12.28) will now be of the form $\exp(ikS^2/2z)\exp(i\mathbf{q}_0\cdot\mathbf{S})$ instead of $\exp(i\mathbf{q}_0\cdot\mathbf{S})$ and eqn (12.29) will have to be adapted accordingly. On illuminating the resulting hologram with plane waves the reconstructed wave front will now contain a term

$$f(S)\exp(-ikS^2/2z)\exp(-i\mathbf{q}_0\cdot\mathbf{S}). \tag{12.35}$$

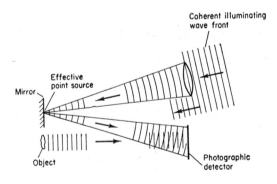

FIG. 12.4. Production of a Fourier Hologram.

The result is as though the wave front is allowed to pass through a convex lens of focal length z as well as through a prism.

There is another way of interpreting eqn (12.35) in the special case that the point source of light and the object are in the same plane during the production of the hologram, as in Fig. 12.4. If $f_0(s)$ is the scattered wave front in a plane just in front of the object, then the wave front described in expression (12.35) above is in fact closely related to the Fourier transform of $f_0(s)$. Thus the point source layout is referred to as Fourier holography, whereas the plane wave front method is called Fresnel holography since they reconstruct respectively the Fraunhofer and Fresnel wave fronts derived from $f_0(s)$.

We can establish the Fourier transform relationship as follows using a discussion analogous to that in Section 11.12. From eqn (11.32) we obtain

$$f(S) = -\frac{ik}{2\pi z} \exp\left[ik(z + S^2/2z)\right] \iint \exp\left(-i\frac{k}{z} s \cdot S\right)$$

$$\times \{f_0(s) \exp(iks^2/2z)\} \, ds. \qquad (12.36)$$

Thus the reconstructed wave front (ignoring the tilt) is given by

$$f(S) \exp(-ikS^2/2z) \propto \iint \exp\left(-i\frac{k}{z} s \cdot S\right)$$

$$\times \{f_0(s) \exp(iks^2/2z)\} \, ds. \qquad (12.37)$$

This represents the Fourier transform of a phase distorted version of $f_0(s)$. This phase distortion is not of great significance since it is equivalent simply to placing a concave lens just in front of the object.

If such a Fourier hologram is reconstructed by illumination with a plane wave front, then it provides a virtual image at infinity. If the hologram is illuminated from a point source at a distance z away then a term $\exp(ikS^2/2z)$ is introduced which cancels the term $\exp(-ikS^2/2z)$ in expression (12.35) and we recover an image of the object in its actual position as usual.

Fourier holograms are useful partly on account of their applications in data processing, to be discussed in the next section, and partly because it can be shown that to obtain a given resolution the photographic emulsion does not have to be so fine grained as in Fresnel holography.

12.9   Optical Processing using Holographic Techniques

In Sections 12.7 and 12.8 we have laid emphasis on 'catching a wave front in mid-air' and reconstructing it to provide a three-dimensional view of the object which gave rise to the wave front. We now shift attention to the problem of processing information which is recorded on photographic transparencies. One problem is that of deconvoluting or 'deblurring' a photograph which has been blurred intentionally or otherwise. If the blurring is intentional then the problem can alternatively be thought of as one of decoding. Another problem is that of correlating two transparencies. If one transparency has a page of written material recorded on it and another transparency just has one word recorded on it then a cross-correlation process applied between the two will result in a transparency with high density at those positions where the chosen word appears on the full page.

These problems have their time-honoured counterparts in signal communication when time (instead of two-dimensional space) is the variable, and the methods adopted in the two subjects are often closely analogous. One technique which we are now to consider is that of Fourier transforming the input data, manipulating or filtering the Fourier transform, and then transforming back again. A pair of convex lenses, each of focal length z, placed a distance $2z$ apart forms the basis for such a process. In Fig. 12.5 we show schematically how a wave front $f(\mathbf{s})$ at the left-hand focal plane is transformed to its Fourier transform $F[(k/z)\mathbf{S}]$ in the central focal plane and is further transformed to give $f(-\mathbf{s})$ at the right-hand focal plane. We have missed out certain proportionality factors and refer the reader to Section 11.12 for a fuller description. Note that the final wave front is inverted since we have used two forward transformations instead of one forward and one inverse transform (see Appendix D). For convenience in the subsequent discussion we will use the vector $\mathbf{q} = k\mathbf{S}/z$ to measure lateral displacements in the central focal plane so that $\mathbf{s}$ and $\mathbf{q}$ are conjugate Fourier variables.

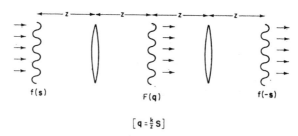

FIG. 12.5. Schematic arrangement for spatial filtering.

If a transparency with amplitude transmission $T(\mathbf{q})$ is placed in the central focal plane, so as to give a wave front $F(\mathbf{q})T(\mathbf{q})$, then at the output in the right-hand focal plane we receive the wave front $f(-\mathbf{s}) \otimes t(-\mathbf{s})$, where

$t(\mathbf{s})$ is the inverse transform of $T(\mathbf{q})$. The problem arises of how to handle a filter function $H(\mathbf{q})$ which is complex in view of the fact that $T(\mathbf{q})$ must necessarily be a real function if we are to rely on photographic transparencies. The answer lies in using a hologram of $H(\mathbf{q})$, or equivalently a Fourier hologram of $h(\mathbf{s})$, as ones filter so that from eqn (12.33)

$$T(\mathbf{q}) = 1 - \frac{\gamma}{2}|H(\mathbf{q})|^2 - \frac{\gamma}{2}H(\mathbf{q})\exp(-i\mathbf{s}_0 \cdot \mathbf{q})$$

$$- \frac{\gamma}{2}H^*(\mathbf{q})\exp(i\mathbf{s}_0 \cdot \mathbf{q}) \qquad (12.38)$$

where γ depends on the photographic emulsion used and $\mathbf{s}_0$ depends on the tilt of the reference beam. The output of the system will now depend on the Fourier transform of the product $F(\mathbf{q})T(\mathbf{q})$ and contains terms giving the convolution and the correlation of $f(\mathbf{s})$ with $h(\mathbf{s})$. In fact Fourier transformation of $F(\pi)T(\pi)$ gives (apart from constant factors, see Section 5.5)

$$\left\{\frac{2}{\gamma}f(-\mathbf{s})\right\} - \{f(-\mathbf{s}) \otimes \rho_h(-\mathbf{s})\} - \{f(-\mathbf{s} - \mathbf{s}_0) \otimes h(-\mathbf{s})\}$$

$$- \{f(-\mathbf{s} + \mathbf{s}_0) * h(-\mathbf{s})\}. \qquad (12.39)$$

Note that provided $f(\mathbf{s})$ is suitably localised the convoluted and correlated outputs are separated, so that we have achieved convolution and correlation as required. The first two terms in expression (12.39) represent a slightly blurred image on the axis of the system, $\rho_h(\mathbf{s})$ being the autocorrelation function of $h(\mathbf{s})$.

The use of a holographic filter has automatically solved for us the problem of cross-correlation. Let us now consider the problem of deconvolution or deblurring. Suppose that our incoming wave front is a convolution of the desired wave front $f(\mathbf{s})$ with a point spread function $h(\mathbf{s})$. In the central focal plane of the arrangement in Fig. 12.5 we thus have the wave front $F(\mathbf{q})H(\mathbf{q})$, and our problem is essentially that of dividing by $H(\mathbf{q})$ to retrieve $F(\mathbf{q})$. An ingenious solution is based on forming a transparency whose amplitude transmission $T_1(\mathbf{q})$ is given by

$$T_1(\mathbf{q}) = \frac{1}{|H(\mathbf{q})|^2}. \qquad (12.40)$$

Such a transparency can be produced by non-holographic means by simply allowing the wave front $h(\mathbf{s})$ to be Fourier transformed by a lens and placing a photographic plate in the appropriate focal plane. The photographic plate receives an intensity $|H(\mathbf{q})|^2$ and according to eqn (12.31) the amplitude transmittance of the resulting negative will be

$$T_1(\mathbf{q}) \propto \{|H(\mathbf{q})|^2\}^{-\gamma/2}. \tag{12.41}$$

Thus the required filter is produced if an emulsion can be found for which $\gamma = 2$. Though difficult such a process can be carried out. The required division is now carried out by noting that

$$\frac{1}{H(\mathbf{q})} = \frac{H^*(\mathbf{q})}{|H(\mathbf{q})|^2}. \tag{12.42}$$

Thus we have only to use the filter described by eqn (12.40) together with a filter $H^*(\mathbf{q})$ to achieve the desired result. In fact the Fourier hologram of $h(\mathbf{s})$ described by eqn (12.38) contains the term $H^*(\mathbf{q})$ so that one simply uses a double filter consisting of this hologram together with the 'division' filter of eqn (12.40).

Chapter 13

X-Ray, Neutron and Electron Diffraction from Stationary Scatterers

13.1 Introduction

Two parallel treatments of the diffraction of waves by three-dimensional systems exist; in one the scattering system is considered as an assembly of point scatterers and in the other it is considered as continuous. Although the two methods can be related (by letting the numbers of point scatterers per unit volume tend to infinity in the one case, and by letting the continuous distribution tend to a set of delta functions in the other) we will nevertheless demonstrate both methods since each appears naturally in different contexts. X-ray diffraction for instance can be based either on a concept of atoms as point scatterers or by treating the electron distribution as continuous. Neutron scattering is more naturally treated using the point scatterer approach, since the scattering nuclei are well localised. Electron diffraction is conveniently discussed using the continuum approach as we shall see. The continuum approach is useful in other problems such as the scattering of light by matter, the atomic spacing being small enough for matter to be considered continuous in this case. Again the scattering of radio waves by atmospheric irregularities lends itself to a continuum approach.

It is our intention to concentrate first on those very general features of diffraction by three-dimensional assemblies which find their expressions in Fourier transforms so that despite the bias indicated by the chapter heading the results will be equally applicable to radio wave, light and acoustic scattering.

13.2   Diffraction by an Assembly of Point Scatterers

Consider first an incident wave

$$\Phi_{\text{inc}} = \phi_0 \cos (\mathbf{k}_0 \cdot \mathbf{r} - \omega_0 t) \tag{13.1}$$

impinging on a single point scatterer at the origin. We assume the scatterer to emit a spherical scattered wave, the amplitude due to the scattered wave at a point $\mathbf{R}$ being (at least for sufficiently large values of $\mathbf{R}$)

$$\Phi_{\text{sc}} = \frac{\phi_0 A}{R} \cos (kR - \omega_0 t + \alpha). \tag{13.2}$$

A is the 'scattering length' determining the strength of the scattering, and α determines that phase of the scattered wave relative to that incident.† A and α depend on the process responsible for scattering and may depend on ω and on the angle of scattering. The inverse dependence on R is consistent with the scattered power (proportional to amplitude squared) obeying an inverse square law. For convenience we will eliminate the time dependence and work with complex amplitudes ϕ defined by

$$\phi_{\text{inc}} = \phi_0 \exp (i\mathbf{k}_0 \cdot \mathbf{r}) \tag{13.3}$$

$$\phi_{\text{sc}} = \frac{\phi_0 a}{R} \exp (ikR) \qquad [a = A \, e^{i\alpha}]. \tag{13.4}$$

The real amplitudes, Φ, may be retrieved as the real part of $\phi \, e^{-i\omega_0 t}$. For Schrödinger waves the wave is itself complex and can be identified directly with $\phi \, e^{-i\omega_0 t}$. Our sign convention is chosen to tally with that universal in wave mechanics, though in classical contexts the equally valid alternative with the time exponent positive and the space exponent negative is found.

For the scattering of X-rays by bound electrons, at frequencies much higher than any atomic resonances, we have $A = -r_e \sin \beta$ and $\alpha = 0$ where r_e ($= 2 \cdot 82 \times 10^{-15}$m) is the 'classical electron radius' and β is the angle between $\mathbf{R}$ and the polarization vector of the incident radiation. The polarisation vector of the scattered radiation is in the direction of $\mathbf{R} \times (\mathbf{n} \times \mathbf{R})$ where $\mathbf{n}$ is the polarization vector of the incident radiation. The negative value of A implies that in forward directions the scattered radiation and incident radiation interfere destructively. For neutron scattering the scattering length may be treated as an empirically determined quantity independent of angle at the wavelengths usually used for diffraction studies. For the treatment of atoms as point scatterers in X-ray or electron

† Values of α other than 0 or π lead to attenuation within a specimen.

diffraction we refer to a discussion on the atomic scattering factor, Section 13.8.

If the scatterer is not at the origin but at some position $\mathbf{r}$ then a phase advance of $(\mathbf{k}_1 - \mathbf{k}_0) \cdot \mathbf{r}$ must be introduced to allow for the altered distances travelled by the incident and scattered waves. $\mathbf{k}_1$ is here a wave vector in the direction of scattering and we have assumed $R \gg r$ so that ray lines from the origin to R and from the scatterer to R may be treated as parallel. The diagram shows the origin of this expression for the path difference.

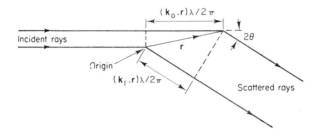

FIGURE 13.1

If we now introduce a scattering vector $\mathbf{K}$ defined by $\mathbf{K} = \mathbf{k}_1 - \mathbf{k}_0$ we may write

$$\phi_{sc} = \frac{\phi_0 a}{R} \exp \{i(kR - \mathbf{K} \cdot \mathbf{r})\}. \tag{13.5}$$

Here again variations in sign convention exist, the scattering vector often being defined as the negative of our $\mathbf{K}$. It is easily verified that $\mathbf{K}$ is a vector normal to the 'mirror plane', the mirror plane being a plane relative to which the incident and scattered rays obey the laws of reflection. The magnitude of $\mathbf{K}$ comes to be $(4\pi \sin \theta)/\lambda$ where 2θ is the angle between $\mathbf{k}_1$ and $\mathbf{k}_0$ (the scattering angle), θ itself being the angle between either ray and the mirror plane (the 'glancing' angle). It is useful to regard the direction of $\mathbf{K}$ as determining indirectly the direction of scattering.

It is now a simple matter to sum up the amplitudes produced at $\mathbf{R}$ by an assembly of N scatterers at positions $\mathbf{r}_i$ ($i = 1, 2, 3, \ldots N$) to give a total amplitude

$$\phi_{sc} = \frac{\phi_0 a}{R} \exp (ikR) \sum_{i=1}^{N} \exp (-i\mathbf{K} \cdot \mathbf{r}_i). \tag{13.6}$$

This expression may be written as a Fourier transform if we introduce the density function $\rho(\mathbf{r}) = \Sigma\delta(\mathbf{r} - \mathbf{r}_i)$ which consists of point delta functions at the positions of the N scatterers. Thus

$$\phi_{sc} = \frac{\phi_0 a}{R} \exp{(ikR)} \int \rho(\mathbf{r}) \exp{(-i\mathbf{K}\cdot\mathbf{r})}\,d\mathbf{r}. \tag{13.7}$$

We have now established the Fourier transform basic to all three-dimensional diffraction problems. In other words we may say that the angular distribution of scattered amplitude depends upon the transform of the 'spatial distribution of scattering centres.' If the scattering centres have different scattering amplitudes, a_i, then clearly we may allow for this by omitting a in eqn (13.7) and redefining $\rho(\mathbf{r})$ as $\Sigma a_i\delta(\mathbf{r} - \mathbf{r}_i)$.

In arriving at eqn (13.7) we have made the important assumption that the scattering is sufficiently weak for the incident wave to progress through the specimen with negligible reduction in intensity, which is equivalent to the assumption of negligible multiple scattering.

13.3   Scattering by a Continuum

We now consider a continuous material of density $\rho(\mathbf{r})$ at $\mathbf{r}$, and assume that an isolated small volume $d\mathbf{r}$ at $\mathbf{r}$ scatters an amount proportional to $\rho(\mathbf{r})\,d\mathbf{r}$ so that we replace eqn (13.5) by the following (the interpretation of 'a' being altered)

$$d\phi_{sc} = \frac{\phi_0 a}{R} \rho(\mathbf{r})\,d\mathbf{r}\exp{\{i(kR - \mathbf{K}\cdot\mathbf{r})\}}. \tag{13.8}$$

An equation of this form arises in many different contexts. One is found when there is a sufficiently high number density of point scatterers per unit volume. Elements of volume may then be chosen small enough so that all the scatterers within it scatter in phase, yet large enough to contain a large number of particles. Another context arises naturally from wave mechanics when the atomic electrons in X-ray diffraction may be taken to have a continuous number density $\rho(\mathbf{r}) = \psi(\mathbf{r})\psi^*(\mathbf{r})$, $\psi(\mathbf{r})$ being the electron wave function within the atom. Another context arises with the weak scattering of of a wave travelling through a medium with non-uniform refractive index; $\rho(\mathbf{r})$ is now equal to $(\mu^2(\mathbf{r}) - 1)$ where $\mu(\mathbf{r})$ is the refractive index at $\mathbf{r}$ and a in eqn (13.8) is put equal to $k_0^2/4\pi$. Finally eqn (13.8) is appropriate to the case of weak electron diffraction or any situation in which a Schrödinger wave is weakly diffracted by a potential $V(\mathbf{r})$. In this case we put $\rho(\mathbf{r}) = V(\mathbf{r})$ and $a = -m/(2\pi\hbar^2)$, m being the mass of the diffracted particle. This approximation is known as the Born approximation.

On integrating eqn (13.8) over all space we obtain

$$\phi_{\text{sc}} = \frac{\phi_0 a}{R} \exp{(ikR)} \int \rho(\mathbf{r}) \exp{(-i\mathbf{K} \cdot \mathbf{r})} \, d\mathbf{r}. \qquad (13.9)$$

This equation is formally identical to eqn (13.7) the only difference lying in the interpretations of the symbols, so that we may in future use the single equation for both situations.

13.4   Cross-Section. Scattering Function

The power flows per unit area in the incident and scattered waves at any point averaged in time are proportional to $|\phi_{\text{inc}}|^2$ and $|\phi_{\text{sc}}|^2$. The power scattered per steradian by the assembly in the direction of a distant point at $\mathbf{R}$ is proportional to $|\phi_{\text{sc}}|^2 R^2$, so that the ratio of power scattered per steradian to incident power per unit area is

$$\sigma(\mathbf{K}) = \frac{|\phi_{\text{sc}}|^2 R^2}{|\phi_{\text{inc}}|^2} = aa^* |\int \rho(\mathbf{r}) \exp{(-i\mathbf{K} \cdot \mathbf{r})} \, d\mathbf{r}|^2. \qquad (13.10)$$

This quantity, $\sigma(\mathbf{K})$, is called the differential scattering cross-section of the whole system. For a single isolated point scatterer we see that $\sigma(\mathbf{K}) = aa^*$. The name arises since the power scattered per steradian is equal to the power incident on an area $\sigma(\mathbf{K})$. For the scattering of X-rays by one single electron this cross-section is $r_e^2 \sin^2 \beta$ (see Section 13.2) for polarized incident radiation. For unpolarized radiation, in which the polarization vector of the incident radiation is constantly changing, then the time averaged cross-section comes to be $r_e^2[1 + \cos^2{(2\theta)}]/2$. The integrated value over all directions of scattering now gives a total (as opposed to differential) scattering cross-section of $8\pi r_e^2/3$.

It is convenient here to define a scattering function $S(\mathbf{K})$ for the system by

$$S(\mathbf{K}) = \frac{|\int \rho(\mathbf{r}) \exp{(-i\mathbf{K} \cdot \mathbf{r})} \, d\mathbf{r}|^2}{\int \rho(\mathbf{r}) \, d\mathbf{r}}. \qquad (13.11)$$

Equation (13.10) may now be rewritten in the form

$$\sigma(\mathbf{K}) = aa^* S(\mathbf{K}) \int \rho(\mathbf{r}) \, d\mathbf{r}$$

and we may interpret the right hand side of this as the product of three parts: aa^* takes account of the physical nature of the scattering process, $\int \rho(\mathbf{r}) \, d\mathbf{r}$ takes into account the amount of scattering material present, and $S(\mathbf{K})$ allows for the effects of interference and depends on the structure of the system. For a system of particles $\int \rho(\mathbf{r}) \, d\mathbf{r} = N$ and $S(\mathbf{K})$ represents the ratio of the intensity actually scattered per particle to what would be scattered by one isolated particle. $S(\mathbf{K})$ is thus greater or less than unity according to whether interference effects are on average constructive or destructive. As we will show later $S(\mathbf{K})$ is unity for a random arrangement of point scatterers. On the continuum model an interpretation of $S(\mathbf{K})$ in words is more cumbersome, though in all cases $S(\mathbf{K})$ contains information on how the angular distribution of the scattered radiation is affected by the purely structural properties of the system.

13.5 Density Autocorrelation Function

In diffraction studies we measure $S(\mathbf{K})$ experimentally and attempt to derive from it information on $\rho(\mathbf{r})$. Two problems arise.

The first problem is that even a complete knowledge of $S(\mathbf{K})$ implies incomplete knowledge of $\rho(\mathbf{r})$ since $S(\mathbf{K})$ is proportional to the spectrum of $\rho(\mathbf{r})$, and thus several different forms for $\rho(\mathbf{r})$ can in principle give the same $S(\mathbf{K})$. Reference to eqn (13.9) shows that if we knew ϕ_{sc} instead of $S(\mathbf{K})$ as a function of $\mathbf{K}$ then we could Fourier invert and obtain $\rho(\mathbf{r})$. However experimentally we measure intensities not amplitudes, and our incomplete information may be traced to a lack of knowledge of the phase of the scattered radiation. It follows from eqn (13.11) and the theorems of Section 5.4 that Fourier inversion of $S(\mathbf{K})$ gives the density autocorrelation function, through the relation

$$S(\mathbf{K}) = \int \exp\left(-i\mathbf{K}\cdot\mathbf{r}\right) \left[\int \rho(\mathbf{r}_1)\rho(\mathbf{r}_1 + \mathbf{r})\, d\mathbf{r}_1\right] d\mathbf{r} / \int \rho(\mathbf{r})\, d\mathbf{r}$$

$$= \int \exp\left(-i\mathbf{K}\cdot\mathbf{r}\right) \langle\rho(\mathbf{r}_1)\rho(\mathbf{r}_1 + \mathbf{r})\rangle\, d\mathbf{r} / \langle\rho(\mathbf{r})\rangle. \tag{13.12}$$

The triangular brackets here refer to averages taken over the whole of the scattering specimen.

The second problem is that a strict Fourier inversion of eqn (13.12) requires a knowledge of $S(\mathbf{K})$ for all values of $\mathbf{K}$. In selecting different values of $\mathbf{K}$ we have at our disposal the possible adjustment of the directions of incidence and scattering relative to the sample and adjustment of the wavelength. Measurement of $S(\mathbf{K})$ for all directions of scattering, with fixed direction of incidence relative to the crystal and fixed wavelength merely covers a range of $\mathbf{K}$ values described by vectors whose tips lie on a sphere, as in Fig. 13.2. Alternatively rotation of the specimen keeping other adjustments fixed can give $\mathbf{K}$ values of fixed magnitude but in any direction relative to the specimen. A combination of the two methods can cover all values of $\mathbf{K}$ with $|K| < 4\pi/\lambda$. If the $S(\mathbf{K})$ is small anyway for larger magnitudes of $\mathbf{K}$ then one has sufficient information to perform the transform, otherwise truncation errors can arise. In an isotropic amorphous material the autocorrelation function of $\rho(\mathbf{r})$ is isotropic and so also therefore is $S(\mathbf{K})$ (see Section 3.6); in this case a measurement of $S(\mathbf{K})$ for all directions of scattering provides sufficient information for the inversion to be carried out, though truncation errors can still occur.

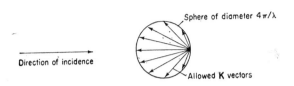

FIGURE 13.2

13.6 The Pair Correlation Function

In a continuous medium eqn (13.12) is useful in that the density auto-correlation function has an easily assimilated meaning. In an assembly of similar point scatterers it is useful to manipulate eqn (13.12) to obtain the following

$$S(\mathbf{K}) = 1 + \int g(\mathbf{r}) \exp(-i\mathbf{K} \cdot \mathbf{r}) \, d\mathbf{r}. \tag{13.13}$$

$g(\mathbf{r})$ is the 'pair correlation function' and is defined as the probability of finding some other particle per unit volume at a distance $\mathbf{r}$ from an arbitrary reference particle. We shall return to a more detailed discussion when we treat diffraction by amorphous materials and liquids, but will include the derivation here since it applies generally to regular as well as irregular arrangements. For an assembly of point scatterers we have:

$$\int \rho(\mathbf{r}_1)\rho(\mathbf{r}_1 + \mathbf{r}) \, d\mathbf{r}_1 = \int \sum_i \delta(\mathbf{r}_1 - \mathbf{r}_i) \sum_j \delta(\mathbf{r}_1 + \mathbf{r} - \mathbf{r}_j) \, d\mathbf{r}_1 \tag{13.14}$$

$$= \sum_i \sum_j \delta(\mathbf{r}_i + \mathbf{r} - \mathbf{r}_j) \tag{13.15}$$

$$= N\delta(\mathbf{r}) + \sum_{i \neq j}\sum \delta(\mathbf{r} - [\mathbf{r}_j - \mathbf{r}_i]) \tag{13.16}$$

$$\approx N\delta(\mathbf{r}) + Ng(\mathbf{r}). \tag{13.17}$$

The second term in eqn (13.16) consists of delta functions at the $N(N-1)$ values of $\mathbf{r}$ corresponding to every interparticle spacing existing in the specimen. Since N is assumed very large in any actual specimen it requires only a small amount of 'blurring' for this array of delta functions to merge together to form the continuous function $Ng(\mathbf{r})$. It is easily verified that this step leads to quite negligible effects. When time effects are included the blurring happens naturally as a result of thermal motion. On inserting eqn (13.17) into eqn (13.12), we arrive at eqn (13.13) as desired. Note that $S(0) = N$ and $\int g(\mathbf{r})\,d\mathbf{r} = N - 1$ as expected, since with the reference particle excluded there are $N - 1$ particles left.

13.7 General Comments

Before proceeding to definite applications of these results it is worth making a few very general observations. We have so far had in mind finite sized specimens. This means that the autocorrelation function of $\rho(\mathbf{r})$ and the pair correlation function $g(\mathbf{r})$ go to zero for values of $\mathbf{r}$ greater than the appropriate dimension of the specimen as a whole. An intense forward diffraction peak always occurs since at angles less than $\theta \approx \lambda/(\text{specimen width})$ all parts of the specimen scatter in phase. For an assembly of particles for instance, $S(\mathbf{K})$ approaches N as $K \to 0$. Our equations give this narrow angle scattering correctly as well as the wider angle scattering associated with microstructure provided $\rho(\mathbf{r})$ and $g(\mathbf{r})$ take account of the finite specimen size as described above. A more subtle difficulty arising from the intense narrow angle scattering lies in the fact that our assumption of weak scattering (made in Section 13.2) can break down. However the fact of its being narrow angle scattering means that even if the incident radiation is quickly attenuated in passing through the specimen it is replaced by the forward scattered radiation. It turns out that the net effect is that of a progressive shift of phase of the forward directed radiation. This may be allowed for by introducing a refractive index and modifying wavelengths within the medium; one may then proceed as though the scattering is weak. The refractive index μ is related to the forward scattering length according to $\mu \approx 1 + \{a\lambda^2 n/2\pi\}$ where n is the mean number of particles per unit volume. If a is complex, implying a phase change on scattering, then we use the real part of a to calculate a refractive index as above giving the ratio of wavelengths outside and inside the specimen. The imaginary part, a_{im}, implies attenuation of the amplitude of radiation by an amount $e^{-\gamma z}$ on passing through a thickness z of material where $\gamma \approx + a_{\text{im}}\lambda n$.

For a randomly arranged set of point scatterers $g(\mathbf{r})$ is clearly a constant for all values of $\mathbf{r}$ small compared to the specimen width so that from eqn (13.13) $S(\mathbf{K}) = 1$ outside the forward scattering peak. Thus interference effects are absent.

For a uniform continuum the autocorrelation function of $\rho(\mathbf{r})$ is constant for $\mathbf{r}$ small compared with the specimen size, so that from eqn (13.12) $S(\mathbf{K})$ falls to zero outside the forward scattering peak. Interference is in this case destructive.

13.8   The Atomic Scattering Factor

If as our 'system' we choose a single atom and consider the Z electrons to be distributed about the centre of charge as origin according to a density function $\rho(\mathbf{r})$ (with $\int \rho(\mathbf{r}) \, d\mathbf{r} = Z$) then it is common in treating X-ray diffraction to define the atomic scattering factor f as:

$$f(\mathbf{K}) = \int \rho(\mathbf{r}) \exp(-i\mathbf{K} \cdot \mathbf{r}) \, d\mathbf{r}. \tag{13.18}$$

Comparison with eqn (13.9) shows that $f(\mathbf{K})$ represents the ratio of amplitude scattered by an atom in a given direction to that scattered by a single electron. The '$S(\mathbf{K})$' of the single atom would be ff^*/Z. $f(\mathbf{K})$ varies from the value Z for small K to zero for large K the changeover being in the region $K \approx 1/(\text{'size of atom'})$. For a spherically symmetric $\rho(\mathbf{r})$ then (see eqn (3.18) and Appendix I) $f(\mathbf{K})$ reduces to

$$f(\mathbf{K}) = \int_0^\infty \frac{\sin Kr}{Kr} 4\pi r^2 \rho(r) \, dr. \tag{13.19}$$

If N atoms are arranged with their centres distributed according to the structural density function

$$\rho_{st}(\mathbf{r}) = \sum_{i=1}^{N} \delta(\mathbf{r} - \mathbf{r}_i)$$

and if each atom has the internal charge density distribution $\rho_{at}(\mathbf{r})$ then the overall density is a convolution of $\rho_{at}(\mathbf{r})$ with $\rho_{st}(\mathbf{r})$. Thus, using the results of eqn (4.31) and (13.11)

$$\rho(\mathbf{r}) = \rho_{at}(\mathbf{r}) \otimes \rho_{st}(\mathbf{r}) \tag{13.20}$$

$$S(\mathbf{K}) = \frac{ff^*}{Z} S_{st}(\mathbf{K}) \tag{13.21}$$

where $S_{st}(\mathbf{K})$ is the scattering function related to $\rho_{st}(\mathbf{r})$ and $S(\mathbf{K})$ is the overall scattering function related to $\rho(\mathbf{r})$. The overall scattering cross-section of the specimen is (from eqns (13.10), (13.11) and (13.21))

$$\sigma(\mathbf{K}) = |a|^2 |f|^2 N S_{st}(\mathbf{K}) \tag{13.22}$$

and we see that the same answer could be obtained by considering the system as an assembly of point scatterers at sites $\mathbf{r}_i$ with a scattering length fa (a being the electron scattering length). It is thus useful to think of the atomic scattering factor as a device for 'reducing' an atom to an equivalent point scatterer.

13.9 Diffraction by a Crystal

Although the overall electron density $\rho(\mathbf{r})$ in a crystal is a complicated function of $\mathbf{r}$ it may be built up (by multiplication and convolution) from simpler functions. We will first show how this may be done and then transform it preparatory to obtaining $S(\mathbf{K})$. The basic feature of a crystal is that a lattice of points exists such that the internal atomic structure of the crystal appears the same relative to each lattice point. The choice of such a set of points is not unique, but given an arbitrary starting point the others follow automatically. Such a lattice may always be described by a density function of the form

$$\rho_{\text{inf}}(\mathbf{r}) = \sum_{n_1} \sum_{n_2} \sum_{n_3} \delta(\mathbf{r} - n_1\mathbf{a} - n_2\mathbf{b} - n_3\mathbf{c}) \qquad (13.23)$$

where $\mathbf{a}$, $\mathbf{b}$ and $\mathbf{c}$ are an appropriately chosen set of vectors known as 'primitive translation vectors'. For a given lattice the choice of $\mathbf{a}$, $\mathbf{b}$ and $\mathbf{c}$ is not quite unique, but this need not concern us here. In the above expression for $\rho_{\text{inf}}(\mathbf{r})$ we consider n_1, n_2 and n_3 to be integers running from $-\infty$ to $+\infty$, and $\rho_{\text{inf}}(\mathbf{r})$ is thus an abstraction describing an infinite lattice. The actual finite lattice associated with a real crystal, $\rho_{\text{cry}}(\mathbf{r})$, may be obtained by limiting the summations in eqn (13.23) or alternatively by truncating $\rho_{\text{inf}}(\mathbf{r})$ by means of a shape function $\rho_{\text{sh}}(\mathbf{r})$ which is unity within the specimen and zero outside. Thus

$$\rho_{\text{cry}}(\mathbf{r}) = \rho_{\text{inf}}(\mathbf{r})\rho_{\text{sh}}(\mathbf{r}). \qquad (13.24)$$

Associated with each lattice point we have a local density $\rho_{\text{loc}}(\mathbf{r})$ which is repeated relative to each point. The overall density $\rho(\mathbf{r})$ is thus a convolution:

$$\rho(\mathbf{r}) = \rho_{\text{cry}}(\mathbf{r}) \otimes \rho_{\text{loc}}(\mathbf{r}). \qquad (13.25)$$

Finally the local distribution may result from an arrangement of N atoms at positions $\mathbf{r}_i$ $(i = 1, \ldots, N)$ relative to the lattice point, the ith atom having an internal electron density $\rho_i(\mathbf{r})$. Thus

$$\rho_{\text{loc}}(\mathbf{r}) = \sum_{i=1}^{N} \rho_i(\mathbf{r} - \mathbf{r}_i). \qquad (13.26)$$

On substituting eqns (13.23), (13.24) and (13.26) into (13.25) and taking the Fourier transform we get

$$\int \exp(-i\mathbf{K} \cdot \mathbf{r}) \rho(\mathbf{r}) \, d\mathbf{r}$$

$$= \frac{F(\mathbf{K})}{|\mathbf{a} \cdot \mathbf{b} \times \mathbf{c}|} \left[W(\mathbf{K}) \otimes \left\{ \sum_{m_1} \sum_{m_2} \sum_{m_3} \delta(\mathbf{K} - m_1\mathbf{a}^* - m_2\mathbf{b}^* - m_2\mathbf{c}^*) \right\} \right]. \qquad (13.27)$$

In transforming $\rho_{\text{loc}}(\mathbf{r})$ we have defined the 'structure factor' $F(\mathbf{K})$ of the unit cell by

$$F(\mathbf{K}) = \int \exp\left(-i\mathbf{K} \cdot \mathbf{r}\right) \rho_{\text{loc}}(\mathbf{r}) \, d\mathbf{r} = \sum_i f_i \exp\left(-i\mathbf{K} \cdot \mathbf{r}_i\right)$$

where f_i is the atomic scattering factor of the ith atom. We have defined, for our present purposes, the width function $W(\mathbf{K})$ by

$$W(\mathbf{K}) = \int \exp\left(-i\mathbf{K} \cdot \mathbf{r}\right) \rho_{\text{sh}}(\mathbf{r}) \, d\mathbf{r}$$

and in transforming $\rho_{\text{inf}}(\mathbf{r})$ we refer to eqn (3.47) where various relationships between $\mathbf{a}^*$, $\mathbf{b}^*$, $\mathbf{c}^*$ and $\mathbf{a}$, $\mathbf{b}$, $\mathbf{c}$ are given. The vectors $\mathbf{a}^*$, $\mathbf{b}^*$, $\mathbf{c}^*$ are the 'reciprocal lattice vectors' and they help to define a 'reciprocal lattice' of points. $|\mathbf{a} \cdot \mathbf{b} \times \mathbf{c}|$ is the volume of the 'unit cell'—a parallelopiped with edges $\mathbf{a}$, $\mathbf{b}$ and $\mathbf{c}$.

Equation (13.27) governs the angular distribution of scattered amplitude (through eqn (13.7)). Let us interpret the terms in it. The diffracted amplitude is strongly concentrated in directions governed by a set of $\mathbf{K}$ vectors whose tips lie on the points of the reciprocal lattice. The actual distribution round these directions is governed by the size and shape of the specimen as a whole through the width function $W(\mathbf{K})$. In most experiments the angular resolution of the apparatus is not good enough to resolve this width. The magnitude of amplitude occurring in any of these well defined diffraction peaks is governed by $F(\mathbf{K})$ which is usually a smoothly and relatively slowly varying function of angle. In summary the lattice vectors determine the positions of the diffraction peaks, the contents of the unit cell determine their strengths, and the specimen shape and size determines their sharpness. If the finite specimen size is alternatively allowed for by letting the sums in eqn (13.23) have the finite limits N_1, N_2 and N_3 (so that there are $N_1 \times N_2 \times N_3$ lattice points in all, centred on the origin) then we may alternatively make use of eqn (3.51) and obtain

$$\int \exp\left(-i\mathbf{K} \cdot \mathbf{r}\right) \rho(\mathbf{r}) \, d\mathbf{r}$$

$$= F(\mathbf{K}) \frac{\sin\left(N_1 \mathbf{K} \cdot \mathbf{a}/2\right)}{\sin\left(\mathbf{K} \cdot \mathbf{a}/2\right)} \frac{\sin\left(N_2 \mathbf{K} \cdot \mathbf{b}/2\right)}{\sin\left(\mathbf{K} \cdot \mathbf{b}/2\right)} \frac{\sin\left(N_3 \mathbf{K} \cdot \mathbf{c}/2\right)}{\sin\left(\mathbf{K} \cdot \mathbf{c}/2\right)}. \quad (13.28)$$

It is easily verified that eqns (13.27) and (13.28) possess the same general features. At a diffraction peak the amplitude rises to $F(\mathbf{K})$ multiplied by the number of lattice points in the specimen, and inside the narrow forward scattering peak the expressions tend towards a value equal to the total number of electrons in the specimen.

From eqns (13.27) and (13.10) we may now derive the cross-section $\sigma(\mathbf{K})$ giving the scattered intensity (see eqn (4.30) for the method of handling the delta functions in eqn (13.27)):

$$\sigma(\mathbf{K}) = \frac{|a|^2 \, |F(\mathbf{K})|^2}{|\mathbf{a} \cdot \mathbf{b} \times \mathbf{c}|^2} \left[|W(\mathbf{K})|^2 \otimes \left\{ \sum_{m_1} \sum_{m_2} \sum_{m_3} \delta(\mathbf{K} - m_1\mathbf{a}^* - m_2\mathbf{b}^* - m_3\mathbf{c}^*) \right\} \right].$$

(13.29)

The alternative based on eqn (13.28) is obvious. $\sigma(\mathbf{K})$ has a peak value, on a diffraction maximum, of $|a|^2 \, |F(\mathbf{K})|^2 N_0^2$ where N_0 is the number of lattice points in the specimen. Bearing in mind that $W(\mathbf{K})$ is concentrated within a small solid angle equal approximately to $\Omega = \pi\lambda^2/(\text{specimen width})^2$, the integrated cross-section over a diffraction peak is of order $|a|^2 \, |F(\mathbf{K})|^2 N_0^2 \Omega$. This can be so large as to invalidate our assumption of weak scattering, and the actual intensity is then less than predicted by eqn (13.29).

13.10   Diffraction by Liquids and Amorphous Materials

Treating the material as an array of point scatterers, and noting that the pair distribution function will be isotropic, we may base our discussion on the following versions of eqn (13.13):

$$S(K) = 1 + \int \exp(-i\mathbf{K} \cdot \mathbf{r}) \, g(r) \, d\mathbf{r}$$

$$= 1 + \int_0^\infty \frac{\sin Kr}{Kr} 4\pi r^2 g(r) \, dr.$$

(13.30)

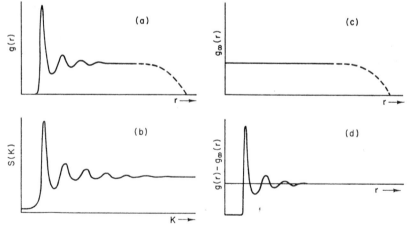

FIGURE 13.3

For a simple monatomic liquid such as argon $g(r)$ has the form shown in Fig. 13.3(a). $g(r)$ is zero up to a value $r \approx D$ (D = atomic diameter) as a result of inpenetrability. The oscillations thereafter imply a certain amount of local order; this order can be thought of as being geometrical in origin and ball bearings packed randomly together show a similar type of order. Beyond the short range order (i.e. beyond $r \approx 5D$) $g(r)$ approaches a uniform value related closely to the mean number density in the specimen. For still larger distances $g(r)$ gradually falls away to zero when $r \approx$ (specimen size), as indicated schematically by the dashed line.

The resultant $S(K)$ is typically as in Fig 13.3(b). The usual intense forward diffraction which occurs at very small angles is not shown. At angles between this and the first broad maximum the small angle scattering is very weak. Then follow broad diffraction maxima in the neighbourhood of $K \approx 1/D$ and at large K, $S(K)$ approaches unity.

In understanding how the main features of $S(K)$ arise it is useful to re-write eqn (13.30) as

$$S(K) = 1 + \int \exp(-i\mathbf{K} \cdot \mathbf{r})\{g(r) - g_\infty(r)\}\, d\mathbf{r} + \int \exp(-i\mathbf{K} \cdot \mathbf{r})\, g_\infty(r)\, d\mathbf{r}. \tag{13.31}$$

$g_\infty(r)$ is a function derived from $g(r)$ by smoothing out the effects of short range order, so that $g_\infty(r)$ is constant for $r \ll$ (specimen size). The actual value of $g_\infty(r)$ for small r is chosen so that $\{g(r) - g_\infty(r)\}$ tends to zero for r greater than the range of short range order; this value of $g_\infty(r)$ for small r is very close to the mean number density of the system. $g_\infty(r)$ and $\{g(r) - g_\infty(r)\}$ are shown in Figs. 13.3(c) and 13.3(d).

The last term in eqn (13.31) gives the strong narrow angle scattering and may be ignored at wider angles. The other term involving $\{g(r) - g_\infty(r)\}$ is responsible for the structure in $S(K)$.

At small angles well inside the first diffraction ring, yet outside the intense forward peak, we get

$$S(K) \approx 1 + \int \{g(r) - g_\infty(r)\}\, d\mathbf{r}. \tag{13.32}$$

The integral in this equation has a value very near to -1 for a tightly packed array of hard spheres, thus explaining the near zero value of $S(K)$ in this region. From Fig. 13.3(d) it can be seen that the negative value of the integral arises chiefly from the dip due to impenetrability at $r < D$. By way of illustration let us consider the opposite extreme; that of a randomly packed arrangement of completely penetrable atoms. In this hypothetical situation two atoms can occupy the same position and $g(r)$ is smooth down to $r = 0$, so that $\{g(r) - g_\infty(r)\}$ is everywhere zero. In this case $S(K) = 1$ at all angles outside the forward peak according to eqn (13.31). It is instructive to look at the value of $\int \{g(r) - g_\infty(r)\}\, d\mathbf{r}$ in a different way. We know that $\int g(r)\, d\mathbf{r} = N - 1$ (see the end of Section 13.6) so it remains to consider $\int g_\infty(r)\, d\mathbf{r}$. Although for $r \ll$ (specimen size) $g_\infty(r)$ is certainly approximately equal to (N/V), where V is the volume of the specimen, we must consider its actual value more carefully. In a tightly packed arrangement of hard spheres, after fixing attention on a reference atom the remaining $N - 1$ are distributed uniformly through the remaining volume, $V - V/N$. Thus

$$g_\infty(0) = (N - 1)/\{V - (V/N)\} = N/V$$

and $\int g_\infty(r)\, d\mathbf{r}$ comes to be N. For the randomly placed assembly of penetrable atoms on the other hand the remaining $N - 1$ atoms are distributed randomly through the whole volume V, so that $g_\infty(0) = (N - 1)/V$ in this case, and $\int g_\infty(r)\, d\mathbf{r} = N - 1$. A similar result applies to a loosely packed array as in a gas). With these rather intuitive geometrical arguments in mind it is not surprising to find that, with some further assumptions based statistical thermodynamics, the narrow angle scattering can be shown to be proportional to the compressibility of the system. In fact it is shown in reference 10 for instance that

$$S(K) = k_B T \beta N/V \qquad [(1/D) \gg K \gg (1/\text{Specimen width})] \qquad (13.33)$$

where k_B is Boltzmann's constant, T the absolute temperature, and β the isothermal compressibility, $(-1/V)(\partial V/\partial p)_T$.

As the specimen size tends to infinity so eqn (13.31) tends more closely to the following form with $g_\infty(r)$ equal to ρ_0 the mean number density, N/V.

$$S(K) = 1 + \int \exp(-i\mathbf{K} \cdot \mathbf{r})\{g(r) - \rho_0\}\, d\mathbf{r} \qquad [K \neq 0].$$

This equation is valid for infinite specimen size (and would remain so with any number substituted for ρ_0) but is not appropriate for handling narrow angle scattering from specimens of finite size.

13.11   Density Perturbations in a Continuous Medium

We consider now a finite scatterer whose density is initially $\rho_0(\mathbf{r})$ and which is perturbed so that the density becomes $\rho(\mathbf{r})$ the perturbation being described by a function $a(\mathbf{r})$ as follows:

$$\rho(\mathbf{r}) = \rho_0(\mathbf{r})\{1 + a(\mathbf{r})\}. \qquad (13.34)$$

The dependence of the scattered amplitude on $\mathbf{K}$ is given by eqn (13.9) and is related to the Fourier transform of $\rho(\mathbf{r})$ which we shall write as $F(\mathbf{K})$. On Fourier transforming eqn (13.34) we obtain, on using the product theorem eqn (3.21),

$$F(\mathbf{K}) = F_0(\mathbf{K}) + \frac{1}{(2\pi)^3} F_0(\mathbf{K}) \otimes A(\mathbf{K}) \qquad (13.35)$$

where

$$F(\mathbf{K}) = \int \exp\left(-i\mathbf{K} \cdot \mathbf{r}\right) \rho(\mathbf{r})\, d\mathbf{r}$$

$$F_0(\mathbf{K}) = \int \exp\left(-i\mathbf{K} \cdot \mathbf{r}\right) \rho_0(\mathbf{r})\, d\mathbf{r}$$

$$A(\mathbf{K}) = \int \exp\left(-i\mathbf{K} \cdot \mathbf{r}\right) a(\mathbf{r})\, d\mathbf{r}.$$

We see from eqn (13.35) that the perturbation has introduced an extra term corresponding to a 'blurred version' of the original function $F_0(\mathbf{K})$. If $\rho_0(\mathbf{r})$ is a rather smooth function with few irregularities or periodicities then the chief feature of $F_0(\mathbf{K})$ is a strong forward peak centred on $\mathbf{K} = 0$, and the $\mathbf{K}$ dependence of the scattering outside this sharp peak is governed essentially by $A(\mathbf{K})$. If $\rho_0(\mathbf{r})$ contains lattice like periodicities then $F_0(\mathbf{K})$ may contain sharp peaks in directions other than $\mathbf{K} = 0$, say at values $\mathbf{K}_i$ of $\mathbf{K}$. In this case the effect of the perturbation is to superpose on these peaks a 'blurred' function of the form $A(\mathbf{K} - \mathbf{K}_i)$.

If we consider the special case that $a(\mathbf{r})$ is a plane harmonic wave of amplitude a_0 phase ϕ and wave vector $\mathbf{Q}$ then we may write

$$a(\mathbf{r}) = a_0 \cos\left\{\mathbf{Q} \cdot \mathbf{r} - \phi\right\}$$

so that, making use of the inverse transformation associated with eqn (3.43), we get:

$$A(\mathbf{K}) = \frac{(2\pi)^3 a_0}{2} \left\{e^{-i\phi}\delta(\mathbf{K} - \mathbf{Q}) + e^{+i\phi}\delta(\mathbf{K} + \mathbf{Q})\right\}$$

and so from eqn (13.35) we obtain:

$$F(\mathbf{K}) = F_0(\mathbf{K}) + \frac{a_0}{2} \left\{e^{-i\phi} F_0(\mathbf{K} - \mathbf{Q}) + e^{+i\phi} F_0(\mathbf{K} + \mathbf{Q})\right\}. \qquad (13.36)$$

The forward scattered peak at $\mathbf{K} = 0$ thus acquires two satellites at $\mathbf{K} = \pm\, \mathbf{Q}$. In order to observe these satellites in an experiment set up with a particular value of $\mathbf{K}$, the perturbing wave must have the right value of $\mathbf{Q}$ and this means in fact that the plane wave fronts must be oriented to act as mirror planes relative to the incident and scattered ray directions and also that the spacing d of these wave fronts must be related to the angle of scattering θ (see Fig. 13.1) and to the wavelength λ of the radiation by:

$$|\mathbf{K}| = |\mathbf{Q}|$$

$$\frac{4\pi \sin \theta}{\lambda} = \frac{2\pi}{d}$$

so that

$$2d \sin \theta = \lambda.$$

This relation is equivalent to the well-known Bragg scattering law.

If $\rho_0(\mathbf{r})$ contains periodicities akin to lattice structure then $F_0(\mathbf{K})$ will possess peaks at, say, values $\mathbf{K}_i$ of $\mathbf{K}$ and the plane wave perturbation will produce extra peaks at $\mathbf{K}_i \pm \mathbf{Q}$, as is clear from eqn (13.36). If $\mathbf{k}_0$ and $\mathbf{k}_1$ are the wave vectors of the incident and scattered radiation then these extra peaks correspond to the condition

$$\mathbf{k}_1 = \mathbf{k}_0 + \mathbf{K}_i + \mathbf{Q}.$$

This relation is interesting because it is identical with a result obtained in a quantum mechanical treatment of the problem when $\mathbf{k}_0$ and $\mathbf{k}_1$ are proportional to the momenta of incoming and outgoing photons, and $\mathbf{K}_i$ and $\mathbf{Q}$ are proportional to momenta exchanged respectively with the lattice and with an excitation or 'phonon'. In the quantum mechanical context this equation appears as a momentum conservation equation. The analogy becomes more striking when a progressive wave is considered (as in Section 14.4) since the frequency changes are then related to energy changes.

If $a(\mathbf{r})$ consists of a distribution of plane waves with amplitudes $z(\mathbf{Q})$ and phases $\phi(\mathbf{Q})$ then we write

$$a(\mathbf{r}) = \int z(\mathbf{Q}) \cos \{\mathbf{Q} \cdot \mathbf{r} - \phi(\mathbf{Q})\} \, d\mathbf{Q}. \tag{13.37}$$

Some choice in interpretation of our symbols arises since a wave with wave vector $\mathbf{Q}$ and phase ϕ is indistinguishable from one with wave vector $-\mathbf{Q}$ and phase $-\phi$. We can either restrict $\mathbf{Q}$ to values whose directions lie in one half of a sphere, or we can allow $\mathbf{Q}$ to assume all directions and put $z(\mathbf{Q}) = z(-\mathbf{Q})$ and $\phi(\mathbf{Q}) = -\phi(-\mathbf{Q})$. We shall adopt the latter course, the integral in eqn (13.37) above covering all values of $\mathbf{Q}$. On account of this symmetry in $z(\mathbf{Q})$ and $\phi(\mathbf{Q})$ we may write eqn (13.37) in the identically equivalent form:

$$a(\mathbf{r}) = \int z(\mathbf{Q}) \exp\left[-i\phi(\mathbf{Q})\right] \exp\left(i\mathbf{Q} \cdot \mathbf{r}\right) d\mathbf{Q}$$

and comparison of this expression with the following inverse transform

$$a(\mathbf{r}) = \frac{1}{(2\pi)^3} \int \exp{(i\mathbf{K} \cdot \mathbf{r})} A(\mathbf{K}) \, d\mathbf{K}$$

shows that

$$A(\mathbf{K}) = (2\pi)^3 z(\mathbf{K}) \exp{[-i\phi(\mathbf{K})]}.$$

Thus the strong forward peak in $F_0(\mathbf{K})$ acquires wings whose $\mathbf{K}$ dependence is given directly by the function $z(\mathbf{K})$.

The analysis given in this section is particularly appropriate to light scattering since the wavelength of light is sufficiently long compared with atomic spacings in solids and liquids for the medium to be approximated to a continuum. Such scattering when it arises from thermally excited waves in liquids or solids is known as Mandelshtam–Brillouin scattering or simply Brillouin scattering. We return to a further discussion when time-dependent perturbations are discussed in Section 14.4.

13.12 Harmonic Distortion of an Assembly of Point Scatterers. Debye–Waller Factor

We consider now an assembly of point scatterers initially at sites $\mathbf{r}_i$ ($i = 1, 2, \ldots, N$) which are displaced on to modified sites at

$$\mathbf{r}_i{}' = \mathbf{r}_i + \mathbf{a}_0 \sin{\{(\mathbf{Q} \cdot \mathbf{r}_i) + \phi\}}.$$

Such a displacement arises for instance when a sound wave passes through a material. Thermal agitation may be represented as a summation of several such waves. If the original density distribution is written as $\rho_0(\mathbf{r})$ and the modified one as $\rho'(\varepsilon)$, then the scattered amplitude is proportional to

$$F'(\mathbf{K}) = \int \exp{(-i\mathbf{K} \cdot \mathbf{r})} \rho'(\mathbf{r}) \, d\mathbf{r} = \sum_{n=-\infty}^{+\infty} e^{in\phi} J_n(-\mathbf{K} \cdot \mathbf{a}_0) F_0(\mathbf{K} - n\mathbf{Q}) \quad (13.38)$$

where

$$F_0(\mathbf{K}) = \int \exp{(-i\mathbf{K} \cdot \mathbf{r})} \rho_0(\mathbf{r}) \, d\mathbf{r}. \quad (13.39)$$

This purely mathematical result is established in Appendix R. It means that a set of functions with amplitudes $J_n(-\mathbf{K} \cdot \mathbf{a}_0)$ and shifted by amounts $n\mathbf{Q}$ are superimposed on the original function, $F_0(\mathbf{K})$. If the original $F_0(\mathbf{K})$ possesses sharp diffraction peaks then we get 'satellites' with intensities proportional to $[J_n(-\mathbf{K} \cdot \mathbf{a}_0)]^2$.

If $(\mathbf{K} \cdot \mathbf{a}_0) \ll 1$ then we may approximate the Bessel functions (see Appendix H) and we find that an unshifted peak and first 'satellite' have intensities equal to $\exp\{-(\mathbf{K}\cdot\mathbf{a}_0)^2/2\}$ and to $(\mathbf{K}\cdot\mathbf{a}_0)^2/4$ respectively expressed as fractions of the original peak intensity.

If we consider the effect of several such disturbances acting simultaneously, the modifications necessary are complicated, except that a simple modification shows an unshifted component to exist with amplitude

$$J_0(\mathbf{K}\cdot\mathbf{a}_0)J_0(\mathbf{K}\cdot\mathbf{a}_1)J_0(\mathbf{K}\cdot\mathbf{a}_2)J_0(\mathbf{K}\cdot\mathbf{a}_3)\ldots J_0(\mathbf{K}\cdot\mathbf{a}_j)\ldots$$

If the displacements are all sufficiently small, and if the original $F_0(\mathbf{K})$ possessed sharp diffraction peaks, then the intensity of the undisplaced peak becomes $\exp\{-\sum_j(\mathbf{K}\cdot\mathbf{a}_j)^2/2\}$. If $\mathbf{u}$ represents the displacement of a particle from equilibrium then

$$\langle(\mathbf{K}\cdot\mathbf{u})^2\rangle = \sum_j (\mathbf{K}\cdot\mathbf{a}_j)^2/2$$

where the triangular brackets represent an average over all particles, so that the intensity may alternatively be written $\exp\{-\langle(\mathbf{K}\cdot\mathbf{u})^2\rangle\}$. This factor governs the reduction in intensity of crystal diffraction peaks with increasing temperature and is known as the Debye–Waller factor. Note that we have established an exponential form only for small displacements.

13.13   Random Displacements of an Assembly of Point Scatterers. Debye–Waller Factor

We assume the scatterer at $\mathbf{r}_i$ to be displaced to a new site at $\mathbf{r}_i + \mathbf{u}_i$. We assume the displacement of an atom to be independent of the displacements of all other atoms, the probability of a displacement into a volume element $d\mathbf{u}$ at $\mathbf{u}$ being $p(\mathbf{u})\,d\mathbf{u}$ with $\int p(\mathbf{u})\,d\mathbf{u} = 1$.

If $S_0(\mathbf{K})$ and $S'(\mathbf{K})$ are the scattering functions of the system in the original and disturbed states respectively, and $g_0(\mathbf{r})$ and $g'(\mathbf{r})$ are the corresponding pair distribution functions, then from eqn (13.13) we have:

$$S_0(\mathbf{K}) = 1 + \int \exp(-i\mathbf{K}\cdot\mathbf{r})\,g_0(\mathbf{r})\,d\mathbf{r} \tag{13.40}$$

$$S'(\mathbf{K}) = 1 + \int \exp(-i\mathbf{K}\cdot\mathbf{r})g'(\mathbf{r})\,d\mathbf{r}. \tag{13.41}$$

In Appendix S we show by a simple physical argument that $g'(\mathbf{r})$ is obtained from $g_0(\mathbf{r})$ and $p(\mathbf{r})$ by a double convolution:

$$g'(\mathbf{r}) = g_0(\mathbf{r}) \otimes p(\mathbf{r}) \otimes p(-\mathbf{r})$$

and it follows from the convolution theorem that eqn (13.41) becomes

$$S'(\mathbf{K}) = 1 + |P(\mathbf{K})|^2 \int \exp(-i\mathbf{K}\cdot\mathbf{r})g_0(\mathbf{r})\,d\mathbf{r} \tag{13.42}$$

where

$$P(\mathbf{K}) = \int \exp(-i\mathbf{K}\cdot\mathbf{r})p(\mathbf{r})\,d\mathbf{r}.$$

Let us interpret this result. $|P(\mathbf{K})|^2$ appears as a Debye–Waller factor. In a perfect crystal we know from our discussion in Section 13.9 that $\int \exp(-i\mathbf{K}\cdot\mathbf{r})g_0(\mathbf{r})\,d\mathbf{r}$ must be very large on a diffraction peak, but must be -1 in between peaks so that $S(\mathbf{K})$ is zero between peaks. The factor $|P(\mathbf{K})|^2$ reduces the intensities of the peaks and also allows a value $S'(\mathbf{K}) = 1 - |P(\mathbf{K})|^2$ in between the peaks. The scattering between peaks is known as 'diffuse scattering' and we see that the difference between the effect of the random displacements considered here and the ordered displacements considered in Section 13.12 lies in the distribution of this diffuse scattering. In Section 13.12 it was concentrated near the diffraction peaks, now it is a broad background falling off with angle.

If the displacements are small and isotropic then $P(\mathbf{K}) \approx 1 - \{\langle(\mathbf{K}\cdot\mathbf{u})^2\rangle/2\}$. This may be seen for instance from the moment expansion of eqn (3.23). Thus

$$|P(\mathbf{K})|^2 \approx \exp\left(-\langle(\mathbf{K}\cdot\mathbf{u})^2\rangle\right)$$

agreeing with the Debye–Waller factor obtained in Section 13.12.

In the special case that $p(\mathbf{u})$ is a Gaussian of width u_0 so that

$$p(\mathbf{u}) = (\pi u_0^2)^{-3/2}\exp(-u^2/u_0^2)$$

then we obtain, on Fourier transforming, the exact relation:

$$|P(\mathbf{K})|^2 = \exp(-K^2 u_0^2/2) = \exp(-K^2\langle u^2\rangle/3) = \exp(-K^2\langle x^2\rangle)$$

$$(13.43)$$

where $\langle u^2\rangle$ and $\langle x^2\rangle$ are the mean square displacement and mean square component of displacement given by

$$\langle u^2\rangle = (3/2)u_0^2$$

$$\langle x^2\rangle = (1/3)\langle u^2\rangle.$$

Chapter 14

X-Ray, Neutron and Electron Diffraction from Time Varying Systems

14.1 Basic Equations

We allow for time variation by modifying the treatment in Sections 13.2 and 13.3, replacing the density function $\rho(\mathbf{r})$ by $\rho(\mathbf{r}, t)$. This applies to a continuous medium or a system of point scatterers and thus covers equally the case of a 'shimmering medium' or of a system of moving particles. The scattered amplitude at a point distant $\mathbf{R}$ from the origin is now written as $\Phi_{sc}(t)$ being, in a classical system, the real part of $\phi_{sc}(t) \exp(-i\omega_0 t)$ where

$$\phi_{sc}(t) = \frac{\phi_0 a}{R} \exp(ikR) \int \exp(-i\mathbf{K}\cdot\mathbf{r}) \rho(\mathbf{r}, t) \, d\mathbf{r}. \tag{14.1}$$

As previously the incident wave, Φ_{inc}, is represented by the real part of $\phi_{inc} e^{-i\omega_0 t}$ where

$$\phi_{inc} = \phi_0 \exp(i\mathbf{k}_0\cdot\mathbf{r}). \tag{14.2}$$

In Appendix T we show that eqn (13.10) becomes generalised to give eqn (14.3) below. We have introduced $\omega = \omega_1 - \omega_0$ to represent the change of angular frequency on scattering where ω_1 is the frequency at which measurements of scattered energy are made.

$$\sigma(\mathbf{K}, \omega) = \frac{aa^*}{2\pi T} \left| \int\int \exp[-i(\mathbf{K}\cdot\mathbf{r} - \omega t)] \rho(\mathbf{r}, t) \, d\mathbf{r} \, dt \right|^2 \tag{14.3}$$

$$= aa^* S(\mathbf{K}, \omega) \frac{1}{T} \int\int \rho(\mathbf{r}, t) \, d\mathbf{r} \, dt \tag{14.4}$$

where

$$S(\mathbf{K}, \omega) = \frac{1}{2\pi} \frac{\left| \iint \exp\left[-i(\mathbf{K} \cdot \mathbf{r} - \omega t)\right] \rho(\mathbf{r}, t) \, d\mathbf{r} \, dt \right|^2}{\iint \rho(\mathbf{r}, t) \, d\mathbf{r} \, dt} \qquad (14.5)$$

$$= \frac{1}{2\pi} \frac{\iint \exp\left[-i(\mathbf{K} \cdot \mathbf{r} - \omega t)\right] \left[\iint \rho(\mathbf{r}_1, t_1) \rho(\mathbf{r}_1 + \mathbf{r}, t_1 + t) \, d\mathbf{r}_1 \, dt_1\right] d\mathbf{r} \, dt}{\iint \rho(\mathbf{r}, t) \, d\mathbf{r} \, dt} \qquad (14.6)$$

$$= \frac{1}{2\pi} \frac{\iint \exp\left[-i(\mathbf{K} \cdot \mathbf{r} - \omega t)\right] \langle \rho(\mathbf{r}_1, t_1) \rho(\mathbf{r}_1 + \mathbf{r}, t_1 + t) \rangle \, d\mathbf{r} \, dt}{\langle \rho(\mathbf{r}_1, t_1) \rangle} . \qquad (14.7)$$

The triangular brackets represent space-time averages over $\mathbf{r}_1$ and t_1. $\sigma(\mathbf{K}, \omega)$, the differential cross-section, represents the ratio of the energy scattered per steradian per unit frequency range by the sample during a time T to the energy incident per unit area during the same time. We shall refer to T as the time duration of the experiment, and shall assume $\rho(\mathbf{r}, t)$ to be zero outside the period $t = 0$ to $t = T$ analogously with $\rho(\mathbf{r}, t)$ being zero for values of $\mathbf{r}$ representing points outside the specimen. $S(\mathbf{K}, \omega)$ is a generalised form of scattering function defined by eqn (14.5); when regarded as a function of $\mathbf{K}$ and ω it is sometimes known as the 'scattering law'. In eqn (14.4) we see that the term $a\,a^*$ is related to the nature of the scattering mechanism,

$$\frac{1}{T} \iint \rho(\mathbf{r}, t) \, d\mathbf{r} \, dt$$

is related to the total amount of scattering material present, and $S(\mathbf{K}, \omega)$ allows for interference and 'Doppler' effects which alter the angular and frequency distribution of scattered energy. For a system of N particles

$$\frac{1}{T} \int \rho(\mathbf{r}, t) \, d\mathbf{r} \, dt = N \qquad (14.8)$$

so that, bearing in mind that the cross-section $\sigma(\mathbf{K})$ for a single stationary particle is aa^* (Section 13.4), we can interpret $S(\mathbf{K}, \omega)$ as the ratio of the energy scattered per steradian per unit frequency range per particle by the whole system to the total energy scattered per steradian by one isolated particle. For a continuum a description of $S(\mathbf{K}, \omega)$ in words is less easy. Equations (14.6) and (14.7) follow mathematically from eqn (14.5) (see

Sections 5.4 and 5.5) and show that $S(\mathbf{K}, \omega)$ is related to the space and time autocorrelation function of $\rho(\mathbf{r}, t)$ by a Fourier transform, a knowledge of this autocorrelation function being the most that can ever be derived from measurements of $S(\mathbf{K}, \omega)$.

14.2 Other Modifications for Time Dependent Systems

Although the motion of the scatterer is taken account of naturally by including ω and t as extra variables complementing $\mathbf{K}$ and $\mathbf{r}$, some associated features are less obvious.

First, it is no longer true to write $K = (4\pi \sin \theta)/\lambda$ (see Section 13.2) as it was in the static case. This arises because the wavelength can change on scattering, so that $k_0 \neq k_1$. Using the cosine rule we get, for a scattering angle 2θ,

$$K^2 = k_0{}^2 + k_1{}^2 - 2 k_0 k_1 \cos 2\theta. \tag{14.9}$$

If a relation between frequency and wavenumber is known, K may be expressed in terms of k_0 and ω. For instance for waves of velocity c we get

$$K^2 = 4\left(k_0{}^2 + \frac{k_0\omega}{c}\right) \sin^2 \theta + \frac{\omega^2}{c^2}. \tag{14.10}$$

For Schrödinger waves with $\omega_{0,1} = \hbar k_{0,1}^2/2m$ we get

$$K^2 = 2 \left\{ k_0{}^2 + \frac{m\omega}{\hbar} - k_0 \left(k_0{}^2 + \frac{2m\omega}{\hbar} \right)^{\frac{1}{2}} \cos 2\theta \right\}. \tag{14.11}$$

In practical cases the frequency changes involved in the scattering of light, X-rays or gamma rays are small enough for eqn (14.10) to reduce to the static value. With neutron scattering however the changes are not negligible and eqn (14.11) cannot then be approximated.

A second point needs to be borne in mind when eqns (14.1)–(14.7) are applied to the scattering of Schrödinger waves as in neutron or electron scattering. We now identify the complex Schrödinger wave function directly with the complex quantities in eqns (14.1), (14.2). The number of particles flowing through unit area per unit time in a Schrödinger wave is given in general by $(\hbar k/m)\psi\psi^*$ where $\psi\psi^*$ represents the probability density of particles and $(\hbar k/m)$ represents their speed (if the wave function is describable locally by a wave number k). Thus eqns (14.3) and (14.4) need to be modified by a factor k_1/k_0 if $\sigma(\mathbf{K}, \omega)$ is to be interpreted as the ratio of the number of particles scattered per steradian per unit frequency range to the number incident per unit area. A further modification is customary since it is more natural to talk of the energy change E rather than the frequency change ω. Introducing an appropriate cross-section, often written as $d^2\sigma/d\Omega\, dE = (k_1/k_0)\, \sigma\, (\mathbf{K}, \omega)\, d\omega/dE$, we get finally

$$\frac{d^2\sigma}{d\Omega\, dE} = \frac{aa^*}{2\pi T} \left(\frac{k_1}{\hbar k_0}\right) \left|\iint \exp\left[-i(\mathbf{K}\cdot\mathbf{r} - \omega t)\right] \rho(\mathbf{r}, t)\, d\mathbf{r}\, dt\right|^2. \quad (14.12)$$

It should be noted that although we have introduced some wave-mechanical ideas by interpreting the scattered waves as Schrödinger waves, we have not developed a fully quantum mechanical treatment since we have ignored the fact that the scattered waves transmit momentum $-\hbar\mathbf{K}$ to the scattering assembly and this can cause local fluctuations in the dynamics of the scatterer. Such a discussion is outside our scope; see for instance, reference 6.

14.3 The Time Dependent Correlation Function—Van Hove's Formula

For a system of particles $\rho(\mathbf{r}, t)$ represents a 'set of delta functions moving about' and the autocorrelation function is difficult to envisage. Instead the correlation function $G(\mathbf{r}, t)$ is useful, eqns (14.5)–(14.7) being alternatively written as

$$S(\mathbf{K}, \omega) = \frac{1}{2\pi} \iint \exp\left[-i(\mathbf{K}\cdot\mathbf{r} - \omega t)\right] G(\mathbf{r}, t)\, d\mathbf{r}\, dt. \quad (14.13)$$

$G(\mathbf{r}, t)\, d\mathbf{r}$ is defined as the probability of finding a particle within an element $d\mathbf{r}$ at $\mathbf{r}$ at time t if there was an arbitrary reference particle at $\mathbf{r} = 0$ at time $t = 0$. The definition does not exclude the reference particle so that, for instance, at time zero we have $G(\mathbf{r}, 0) = \delta(\mathbf{r}) + \overline{g(\mathbf{r})}$ where $\overline{g(\mathbf{r})}$ is the time average of the ordinary pair correlation function (Section 13.6). In a liquid the evolution of $G(\mathbf{r}, t)$ with time is shown schematically in Figs. 14.1a, b, c, and d, at various times t_1, t_2 and t_3. The central delta function spreads out gradually with time as a result of diffusive motion, and the partial order in the positions of near neighbours gradually smooths out until, at times long compared with some relaxation time, $G(\mathbf{r}, t)$ approaches a uniform value equal to the mean number density of the system. The dashed portions illustrate that $G(\mathbf{r}, t)$ goes to zero outside the specimen.

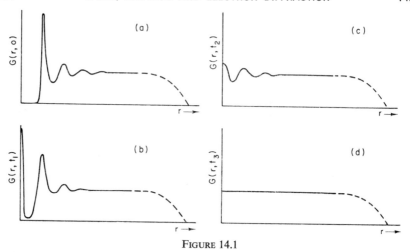

FIGURE 14.1

The equivalence between eqns (14.6) and (14.13) may be shown as follows. In a system of N particles we let $\mathbf{r}_i(t)$ be the position of the ith particle at time t.

$$\iint \rho(\mathbf{r}_1, t_1)\, \rho(\mathbf{r}_1 + \mathbf{r}, t_1 + t)\, \mathrm{d}\mathbf{r}_1\, \mathrm{d}t_1$$

$$= \iint \sum_i \delta\{\mathbf{r}_1 - \mathbf{r}_i(t_1)\} \sum_j \delta\{\mathbf{r}_1 + \mathbf{r} - \mathbf{r}_j(t_1 + t)\}\, \mathrm{d}\mathbf{r}_1\, \mathrm{d}t_1 \qquad (14.14)$$

$$= \int \sum_i \sum_j \delta\{\mathbf{r}_i(t_1) + \mathbf{r} - \mathbf{r}_j(t_1 + t)\}\, \mathrm{d}t_1 \qquad (14.15)$$

$$= \int N G(\mathbf{r}, t, t_1)\, \mathrm{d}t_1 \qquad (14.16)$$

$$= N T G(\mathbf{r}, t). \qquad (14.17)$$

Equation (14.15) follows from (14.14) if use is made of the three-dimensional analogue of eqn (B.9). The summation in eqn (14.15) is a function of $\mathbf{r}, t$ and t_1 which for given values of t and t_1 consists of a set of N^2 delta functions at values of $\mathbf{r}$ corresponding to the set of N^2 distances between the position of the ith particle at time t_1 and the jth particle at time $t_1 + t$ (for all i and j). Since the number of particles is large we need only a little 'blurring' (i.e. convolution with some broadening function) for the summation to become $NG(\mathbf{r}_1 t, t_1)$ where $G(\mathbf{r}, t, t_1)$ is the probability of finding a particle at $\mathbf{r}$ at time $t_1 + t$ if there was a reference particle at $\mathbf{r} = 0$ at time t_1. If we assume $G(\mathbf{r}, t, t_1)$ is independent of t_1 (provided t_1 lies within 0 and T, that is during the duration of the experiment) then eqn (14.17) follows from eqn (14.16). Equation (14.13) now follows from eqns (14.6) and 14.17) if we realize that $\iint \rho(\mathbf{r}, t) \, d\mathbf{r} \, dt = NT$.

$G(\mathbf{r}, t)$ and $S(\mathbf{K}, \omega)$ are often divided into two parts

$$
\left.
\begin{aligned}
G(\mathbf{r}, t) &= G_s(\mathbf{r}, t) + G_d(\mathbf{r}, t) \\
S(\mathbf{K}, \omega) &= S_s(\mathbf{K}, \omega) + S_d(\mathbf{K}, \omega).
\end{aligned}
\right\} \quad (14.18)
$$

$G_s(\mathbf{r}, t)$ and $G_d(\mathbf{r}, t)$ are called the self and distinct correlation functions respectively, and if an arbitrary reference particle is at $\mathbf{r} = 0$ at $t = 0$ then the former function represents the probability of finding the *same* reference particle at position $\mathbf{r}$ at time t, the latter function representing the probability of finding some *other* particle there. Clearly $G_s(\mathbf{r}, 0) = \delta(\mathbf{r})$ and $G_d(\mathbf{r}, 0) = \overline{g(\mathbf{r})}$ where the bar once again represents a time average. $S_s(\mathbf{K}, \omega)$ and $S_d(\mathbf{K}, \omega)$ are defined from $G_s(\mathbf{r}, t)$ and $G_d(\mathbf{r}, t)$ through equations analogous to (14.13).

If, as in X-ray work, we are unable to resolve the frequency changes, then our measuring apparatus measures $\int S(\mathbf{K}, \omega) \, d\omega$. It is satisfying that we are able to retrieve the 'static' formula eqn (13.13) on performing this integration. We get from eqn (14.13)

$$
\int S(\mathbf{K}, \omega) \, d\omega = \frac{1}{2\pi} \iiint \exp\left[-i(\mathbf{K} \cdot \mathbf{r} - \omega t)\right] G(\mathbf{r}, t) \, d\mathbf{r} \, dt \, d\omega \quad (14.19)
$$

$$
= \int \exp(-i\mathbf{K} \cdot \mathbf{r}) \, G(\mathbf{r}, 0) \, d\mathbf{r} \quad (14.20)
$$

$$
= \int \exp(-i\mathbf{K} \cdot \mathbf{r}) \{\delta(\mathbf{r}) + \overline{g(\mathbf{r})}\} \, d\mathbf{r} \quad (14.21)
$$

$$
= 1 + \int \exp(-i\mathbf{K} \cdot \mathbf{r}) \overline{g(\mathbf{r})} \, d\mathbf{r} \quad (14.22)
$$

$$
= S(\mathbf{K}). \quad (14.23)
$$

The integration leading from eqn (14.19) to eqn (14.20) is explained in Appendix B, eqn (B.11). Thus the information obtained from such X-ray work relates to the instantaneous spatial arrangement inside the specimen averaged in time, and the results of Chapter 13 remain useful even if the system is not in fact static.

Equation (14.13) was first developed in the context of neutron scattering by Van Hove, who showed that a similar expression was applicable in a completely quantum mechanical situation when for instance the scattering particle may not have a well localised wave function and when the effects of recoil should be included. In such situations $G(\mathbf{r}, t)$ is defined using quantum mechanical operators and can be a complex quantity, the present interpretation then appears as a classical limit. Note that $S(\mathbf{K}, \omega)$ is in all cases real and this means that $G(\mathbf{r}, t) = G(-\mathbf{r}, -t)$ a result which a little thought will show to be physically justifiable in its own right. The effects of recoil would mean that $S(\mathbf{K}, \omega) \neq S(-\mathbf{K}, -\omega)$ since a loss of energy on scattering might for instance become more likely than a gain; this immediately means that if an equation of the form of (14.13) is to apply then $G(\mathbf{r}, t)$ must be complex since real functions only have a real transform if they are even functions (see Section 2.3).

14.4 Density Perturbations in a Continuous Medium—Brillouin Scattering

The treatment of Section 13.11 may readily be adapted to the time dependent case. We write the density within a finite scattering sample as

$$\rho(\mathbf{r}, t) = \rho_0(\mathbf{r}, t) \{1 + a(\mathbf{r}, t)\} \tag{14.24}$$

where $\rho_0(\mathbf{r}, t)$ is the original density and $a(\mathbf{r}, t)$ represents the perturbation. Often it is convenient to consider the original density as time independent, but our treatment allows it to be time dependent.

The $\mathbf{K}$ and ω dependence of the scattered intensity is given by eqn (14.3) and is related to the Fourier transform of $\rho(\mathbf{r}, t)$ which we shall write as $F(\mathbf{K}, \omega)$. On transforming eqn (14.24) we get the following, involving a convolution over $\mathbf{K}$ and ω.

$$F(\mathbf{K}, \omega) = F_0(\mathbf{K}, \omega) + \frac{1}{(2\pi)^4} F_0(\mathbf{K}, \omega) \otimes A(\mathbf{K}, \omega) \tag{14.25}$$

where

$$F(\mathbf{K}, \omega) = \int \exp\left[-i(\mathbf{K}\cdot\mathbf{r} - \omega t)\right] \rho(\mathbf{r}, t) \, d\mathbf{r} \, dt$$

$$F_0(\mathbf{K}, \omega) = \int \exp\left[-i(\mathbf{K}\cdot\mathbf{r} - \omega t)\right] \rho_0(\mathbf{r}, t) \, d\mathbf{r} \, dt$$

$$A(\mathbf{K}, \omega) = \int \exp\left[-i(\mathbf{K}\cdot\mathbf{r} - \omega t)\right] a(\mathbf{r}, t) \, d\mathbf{r} \, dt.$$

If we consider a disturbance consisting of a travelling plane wave, then we may write

$$a(\mathbf{r}, t) = a_0 \cos\{\mathbf{Q}\cdot\mathbf{r} - \Omega t\}. \tag{14.26}$$

On Fourier transforming this over space and time variables we obtain

$$A(\mathbf{K}, \omega) = \frac{(2\pi)^4 a_0}{2} \left\{ \delta(\mathbf{K} - \mathbf{Q}) \delta(\omega - \Omega) + \delta(\mathbf{K} + \mathbf{Q}) \delta(\omega + \Omega) \right\} \quad (14.27)$$

so that from eqn (14.25) we obtain

$$F(\mathbf{K}, \omega) = F_0(\mathbf{K}, \omega) + \frac{a_0}{2} \left\{ F_0(\mathbf{K} - \mathbf{Q}, \omega - \Omega) + F_0(\mathbf{K} + \mathbf{Q}, \omega + \Omega) \right\}.$$

$$(14.28)$$

Thus if $F_0(\mathbf{K}, \omega)$ possesses an intense elastic forward scattered peak with $\mathbf{K} = 0$ and $\omega = 0$ then new peaks are produced at $\mathbf{K} = +\mathbf{Q}$ and at $\mathbf{K} = -\mathbf{Q}$ the scattered radiation being respectively high and low in frequency compared with the incident frequency by amounts equal to Ω. As was shown in Section 13.11 the condition $\mathbf{K} = \mathbf{Q}$ is similar to Bragg's diffraction formula. The new condition $\omega = \pm \Omega$ can be interpreted as a Doppler shift due to reflection off a moving mirror plane, the plus or minus sign depending on the direction of motion of the wave relative to $\mathbf{K}$. If the speeds of the incident radiation and of the disturbing wave are c and v respectively, and if ω_0 and $\mathbf{k}_0$ are the frequency and wave vector of the incident radiation then the fractional change in frequency is given by

$$\frac{\Omega}{\omega_0} = \frac{|\mathbf{K}|v}{|\mathbf{k}_0|c} = \frac{2v \sin \theta}{c} \quad (14.29)$$

where 2θ is the angle of scattering as in Fig. 13.1 and $|\mathbf{K}|$ and $|\mathbf{k}_0|$ are related as in Section 13.2. This is just as would be expected from a Doppler shift.

If $\rho_0(\mathbf{r}, t)$ is independent of time and contains spatial periodicities due to lattice structure then $F_0(\mathbf{K}, \omega)$ will contain sharp elastic peaks (i.e. $\omega = 0$) at certain values of $\mathbf{K}$, say at $\mathbf{K}_i$. In this case the progressive plane wave disturbance will give rise to sharp peaks at values $\mathbf{K} = \mathbf{K}_i \pm \mathbf{Q}$ with frequency changes $\omega = \pm \Omega$. Using ω_0 and $\mathbf{k}_0$ to refer to the frequency and wave vector of the incident radiation, and ω_1 and $\mathbf{k}_1$ to refer correspondingly to the scattered radiation we may describe the condition for strong scattering by the two equations

$$\left. \begin{array}{l} \mathbf{k}_1 = \mathbf{k}_0 + \mathbf{K}_i \pm \mathbf{Q} \\[2mm] \omega_1 = \omega_0 \pm \Omega. \end{array} \right\} \quad (14.30)$$

In a quantum mechanical treatment these equations remain true, representing respectively the conservation of momentum and energy, $\mathbf{Q}$ and Ω being proportional to the momentum and energy given up by the excitation to the scattered radiation.

Let us now consider a continuous distribution of plane waves so that

$$a(\mathbf{r}, t) = \int z(\mathbf{Q}) \cos \{\mathbf{Q} \cdot \mathbf{r} - \Omega(\mathbf{Q})t\} \, d\mathbf{Q}. \tag{14.31}$$

The function $z(\mathbf{Q})$ is now the amplitude spectrum of the waves, and $\Omega(\mathbf{Q})$ is the dispersion relation of the medium relating frequency to wave vector. For the special case that waves of all wavelengths travel at the same speed v then $\Omega(\mathbf{Q})$ has the simple form $\Omega = v|\mathbf{Q}|$. We adopt the convention that Ω is necessarily positive so that $\mathbf{Q}$ points in the direction of propagation of a wave. Often it will be true that $z(\mathbf{Q}) = z(-\mathbf{Q})$ and $\Omega(\mathbf{Q}) = \Omega(-\mathbf{Q})$, but these relations will not necessarily hold, the first equation being untrue if the wave system is not isotropic and the second equation being untrue if the medium is itself in motion parallel to $\mathbf{Q}$. The Fourier transform of eqn (14.31) may be obtained by generalising the expression for $A(\mathbf{K}, \omega)$ given in eqn (14.27). We replace a_0 by $z(\mathbf{Q})$, replace Ω by $\Omega(\mathbf{Q})$ and integrate over all values of $\mathbf{Q}$ giving

$$A(\mathbf{K}, \omega) = \frac{(2\pi)^4}{2} \int z(\mathbf{Q}) \left[\delta(\mathbf{K} - \mathbf{Q}) \delta\{\omega - \Omega(\mathbf{Q})\} \right.$$

$$\left. + \delta(\mathbf{K} + \mathbf{Q}) \delta\{\omega + \Omega(\mathbf{Q})\} \right] d\mathbf{Q}$$

$$= \frac{(2\pi)^4}{2} \left[z(\mathbf{K}) \delta\{\omega - \Omega(\mathbf{K})\} + z(-\mathbf{K}) \delta\{\omega + \Omega(-\mathbf{K})\} \right]. \tag{14.32}$$

If the chief feature of $F_0(\mathbf{K}, \omega)$ is an elastic forward scattered peak then the scattering amplitude at some value of $\mathbf{K}$ outside this peak will be governed by $A(\mathbf{K}, \omega)$ and we see from eqn (14.32) that two frequency components appear, shifted up and down in frequency by amounts $\Omega(\mathbf{Q})$ and $\Omega(-\mathbf{Q})$ and having amplitudes respectively proportional to $z(\mathbf{Q})$ and $z(-\mathbf{Q})$. Such components may be measured in the scattering of light from solids and liquids and they are known as the Mandelshtam–Brillouin doublet or simply Brillouin doublet. Frequency broadening of these components may be attributed to scattering from plane waves which are damped and have a finite range and lifetime so that a spread of frequencies is associated with each value of Ω. A third component centered on zero frequency change and having a finite spread of frequencies is often observed also, and this may be attributed to scattering from random fluctuations in density. This is sometimes referred to as the Rayleigh component, and its intensity $|A(\mathbf{K}, \omega)|^2$ may be related to the Fourier transform of the space and time autocorrelation function of $a(\mathbf{r}, t)$.

14.5 Harmonic Distortion of an Assembly of Point Scatterers— Debye–Waller Factor

We can generalize the discussion of Section 13.12 to allow for a moving assembly of points. We consider scatterers whose original motion corresponds to a set of time dependent sites $r_i(t)$ which, as a result of a wave moving through the assembly, are changed to modified sites at $r_i'(t) = r_i(t) + a_0 \sin \{(Q \cdot r_i(t)) - \Omega t + \phi\}$. The original and modified states may be described by density functions $\rho_0(r, t)$ and $\rho'(r, t)$ which consist of delta functions at the appropriate sites. The scattered amplitudes from the original and modified states will depend upon $F_0(K, \omega)$ and $F'(K, \omega)$ defined as below, and the relationship between them is given in eqn (14.35).

$$F_0(K, \omega) = \int \exp\left[-i(K \cdot r - \omega t)\right] \rho_0(r, t) \, dr \, dt \qquad (14.33)$$

$$F'(K, \omega) = \int \exp\left[-i(K \cdot r - \omega t)\right] \rho'(r, t) \, dr \, dt \qquad (14.34)$$

$$= \sum_{n=-\infty}^{+\infty} J_n(-K \cdot a_0) e^{in\phi} F_0(K - nQ, \omega - n\Omega). \qquad (14.35)$$

Equation (14.35) can be thought of simply as the four-dimensional analogue of eqn (13.38). It is established more rigorously in Appendix R, (eqn (R.25)).

We may interpret this result to mean that superimposed on the original function $F_0(K, \omega)$ we have replicas which are shifted both in K value and in frequency, and with amplitudes falling off as $J_n(-K \cdot a_0)$. If the 'original' state was a static one, with moreover a crystalline regularity so that $F_0(K, \omega)$ could be written $F_0(K) \delta(\omega)$ where $F_0(K)$ corresponded to sharp peaks in certain directions, then these peaks will acquire satellites at positions $K_i \pm nQ$ relative to a peak at K_i, each peak corresponding to a frequency shift $\pm n\Omega$, and having an intensity proportional to $J_n^2(-K \cdot a_0)$. If the specimen is non-crystalline then $F_0(K)$ will not possess sharp peaks, and measurements of scattering intensity versus angle will not distinguish the unmodified component from the superimposed satellites. However if, as in neutron or Mössbauer scattering, one has sufficient energy resolution to resolve the elastic scattering from the inelastic one will find that the elastic component has an intensity governed by a Debye–Waller factor. This can be seen by writing the structure factor $S'(K, \omega)$ proportional to $|F'(K, \omega)|^2$ as follows, in terms of the static structure factor $S_0(K)$,

$$S'(K, \omega) = J_0^2(-K \cdot a_0) S_0(K) \delta(\omega) + \text{Frequency shifted terms}. \qquad (14.36)$$

If, as in the thermal excitation of a specimen, one has a superposition of many disturbing waves then the frequency shifted terms become complicated, but the intensity of the elastic component comes to be a product of Bessel functions which for small vibration amplitudes can be approximated just as in Section 13.12.

14.6 Independent Vibrations of an Assembly of Point Scatterers

We consider, as the time dependent analogue of the case covered in Section 13.13, an assembly of point scatterers vibrating about fixed sites. The vibrations are independent but similar statistically so that the probability of a particle having moved a distance $\mathbf{r}$ in time t is the same for all particles and can be put as $G_s(\mathbf{r}, t)$. Note that this is measured from an arbitrary starting instant so $\mathbf{r}$ is not to be interpreted as the displacement from the site.

We may write

$$G(\mathbf{r}, t) = G_s(\mathbf{r}, t) + \{G(\mathbf{r}, 0) - \delta(\mathbf{r})\}. \qquad (14.37)$$

In this equation we have separated off the contribution from the self correlation function, and have expressed the fact that the distinct part is independent of time and thus equal to $\{G(\mathbf{r}, 0) - \delta(\mathbf{r})\}$, its value at $t = 0$. This fact can be seen to be true by considering that since the motions are independent the distribution of other atoms relative to a reference atom will on average appear the same at $t = 0$ as at any other time.

On Fourier transforming 14.37 to get the scattering law we obtain

$$S(\mathbf{K}, \omega) = S_s(\mathbf{K}, \omega) + \{S(\mathbf{K}) - 1\} \delta(\omega) \qquad (14.38)$$

where $S(\mathbf{K})$ is the integrated value of $S(\mathbf{K})$ over all frequencies (see eqns (14.20) to (14.23)).

$S_s(\mathbf{K}, \omega)$ may be split into an elastic and an inelastic part for a bound motion such as vibration; this is discussed in Section 14.7 and using eqn (14.43) we may rewrite eqn (14.38) as

$$S(\mathbf{K}, \omega) = \delta(\omega) [S(\mathbf{K}) - \{1 - D(\mathbf{K})\}] + \{1 - D(\mathbf{K})\} f(\mathbf{K}, \omega). \qquad (14.39)$$

$D(\mathbf{K})$ is here used as in Section 14.7 to represent the Debye–Waller factor, and $f(\mathbf{K}, \omega)$ is a function introduced to represent the inelastic scattering and normalised so that $\int f(\mathbf{K}, \omega) \, d\omega = 1$.

Remembering from Section 13.13 that $S(\mathbf{K})$ may itself be related to the scattering law $S_0(\mathbf{K})$ that would apply in the absence of displacement from the reference sites, we get combining eqns (14.39) and (13.42) and allowing for the different notations,

$$S(\mathbf{K}, \omega) = \delta(\omega)\, D(\mathbf{K})\, S_0(\mathbf{K}) + \{1 - D(\mathbf{K})\} f(\mathbf{K}, \omega). \tag{14.40}$$

Equation (14.40) shows that the effect of vibration in an ideal crystal is to reduce the intensities of the main peaks by a Debye–Waller factor leaving these peaks elastic, and to create a diffuse background which is inelastic. Equation (14.39) shows that with an amorphous material, for any value of $\mathbf{K}$ we have some elastic and some inelastic scattering the ratio of intensities being $[S(\mathbf{K}) - \{1 - D(\mathbf{K})\}] : [1 - D(\mathbf{K})]$. The proportion of elastic scattering is thus higher nearer the peak values of $S(\mathbf{K})$.

14.7 The Self Correlation Function for a Bound Motion

As a first step to understanding the total correlation function it is useful to consider that contribution from the self correlation function. A common special situation is that of a bound motion such as vibration or rotation about a fixed site, the basic feature being that $G(\mathbf{r}, t)$ tends at large times to a form still peaked about the origin. We will write this limiting form as $G_s(\mathbf{r}, \infty)$, though it is understood that we mean a time long on a microscopic scale but short compared with the duration of the experiment. We may now write $G_s(\mathbf{r}, t)$ as a time independent and a time dependent part thus

$$G_s(\mathbf{r}, t) = G_s(\mathbf{r}, \infty) + \{G_s(\mathbf{r}, t) - G_s(\mathbf{r}, \infty)\} \tag{14.41}$$

and on transforming obtain

$$S_s(\mathbf{K}, \omega) = \delta(\omega) \int \exp(-i\mathbf{K}\cdot\mathbf{r})\, G_s(\mathbf{r}, \infty)\, d\mathbf{r}\, dt$$

$$+ \frac{1}{2\pi} \int \exp[-i(\mathbf{K}\cdot\mathbf{r} - \omega t)]\, [G(\mathbf{r}, t) - G_s(\mathbf{r}, \infty)]\, d\mathbf{r}\, dt. \tag{14.42}$$

The first term represents elastic scattering, and the second inelastic scattering. The intensity of the elastic scattering may be represented as a Debye–Waller factor $D(\mathbf{K})$ and we then obtain

$$S_s(\mathbf{K}, \omega) = D(\mathbf{K})\, \delta(\omega) + \{1 - D(\mathbf{K})\} f(\mathbf{K}, \omega) \tag{14.43}$$

$$D(\mathbf{K}) = \int \exp(-i\,\mathbf{K}\cdot\mathbf{r})\, G_s(\mathbf{r}, \infty)\, d\mathbf{r} \tag{14.44}$$

where $f(\mathbf{K}, \omega)$ is introduced to represent the inelastic scattering and is normalised so that $\int f(\mathbf{K}, \omega)\, d\omega = 1$. The term $\{1 - D(\mathbf{K})\}$ arises since we know that $\int S_s(\mathbf{K}, \omega)\, d\omega = 1$ (see eqns (14.19) to (14.22)).

This rather general expression for the Debye–Waller factor is equivalent to the expression given earlier (following eqn (13.42). The equivalence depends on the equality

$$G_s(\mathbf{r}, \infty) = p(\mathbf{r}) \otimes p(-\mathbf{r})$$

so that $D(\mathbf{K}) = |P(\mathbf{K})|^2$, where $p(\mathbf{r})$ and $P(\mathbf{K})$ have their previous meanings. We will not prove the equivalence but will point out that $G_s(\mathbf{r}, \infty)$ refers to the probability of travelling a distance $\mathbf{r}$ in infinite time from an arbitrary starting time, whereas $p(\mathbf{r})$ refers to the probability of a displacement $\mathbf{r}$ from the centre of motion. The proof follows much the same lines as that given in Appendix S.

If $G_s(\mathbf{r}, \infty)$ has the special form below, as is the case for instance in the Debye model of a crystal, then we can use a standard transform to obtain an explicit expression for $D(\mathbf{K})$ thus:

$$G_s(\mathbf{r}, \infty) = \left(2\pi r^2(\infty)/3\right)^{-3/2} \exp\left(-3r^2/2r^2(\infty)\right) \tag{14.45}$$

$$D(\mathbf{K}) = \exp\left(-K^2 r^2(\infty)/6\right) = \exp\left(-K^2 x^2(\infty)/2\right)$$

$$= \exp\left(-K^2 \langle x^2 \rangle\right). \tag{14.46}$$

In these equations $r^2(\infty)$ is the mean square distance moved in infinite time, $x^2(\infty)$ is the mean square component of distance moved along a fixed axis, and $\langle x^2 \rangle$ is the mean square x-displacement from the centre of motion.

14.8   Unbound Motion—Diffusion

For an unbound motion $G_s(\mathbf{r}, \infty)$ is just a uniform function within the specimen equal to the mean number density, and it follows from the above discussion that the only elastic scattering which occurs is within the narrow forward scattered peak. As a simple special case consider a classical diffusive motion for the scattering particles. This is a motion for which:

$$G_s(\mathbf{r}, t) = (4\pi D |t|)^{-3/2} \exp\left(-r^2/4D |t|\right) \tag{14.47}$$

$$r^2(t) = 3x^2(t) = 6D |t| \tag{14.48}$$

where D is the self diffusion coefficient and $r^2(t)$ and $x^2(t)$ are the mean square distance and mean square component of distance travelled in time t.

On Fourier transforming using eqns (3.38) and (2.39) we get:

$$S_s(\mathbf{K}, \omega) = \frac{1}{\pi} \frac{DK^2}{\omega^2 + D^2 K^4}. \tag{14.49}$$

It remains to consider the form of $G(\mathbf{r}, t)$ as opposed to $G_s(\mathbf{r}, t)$. No simple analytical expression exists to cover a system of closely packed particles such as a liquid. An approximation which reproduces some features of $G(\mathbf{r}, t)$ correctly is the following:

$$G(\mathbf{r}, t) \approx G(\mathbf{r}, 0) \otimes G_s(\mathbf{r}, t). \tag{14.50}$$

It is intuitively safe to guess that $G(\mathbf{r}, t)$, which starts off in time with the form $G(\mathbf{r}, 0)$, evolves by letting the peaks and dips gradually smooth out as shown diagramatically in Fig. 14.1. Regarding $G_s(\mathbf{r}, t)$ as a 'broadening' function, the convolution in eqn (14.50) provides just this effect, the time taken for most of the smoothing to be accomplished being equal roughly to the time taken for a particle to move a distance equal to the particle diameter. Such an approximation does not take account of the detailed correlations in the motions of neighbouring particles which must exist if they are not to interpenetrate one another. On Fourier transforming eqn (14.50) we obtain (using eqns (14.20)–(14.23))

$$S(\mathbf{K}, \omega) \approx S(\mathbf{K}) \, S_s(\mathbf{K}, \omega) \tag{14.51}$$

showing that in this approximation eqn (14.49) gives the energy distribution of scattering correctly.

Appendices

Appendix A—Evaluation of Fourier Series Coefficients

The derivation is most succinctly written for the complex exponential expansion, from which the other results may of course be derived. The derivation takes just four lines as below. Note that we use $\oint$ to represent $\int_{t_1}^{t_1+\tau}$ where τ is the repetition period, and t_1 is arbitrary; we put $\omega = 2\pi/\tau$. The relation (A.3) is a straightforward result of integration and may be checked by writing out the complex exponential in terms of a sine and cosine.

If
$$f(t) = \sum_{n=-\infty}^{+\infty} g_n e^{+in\omega t} \tag{A.1}$$

then
$$\oint f(t) e^{-im\omega t} \, dt = \oint \left\{ \sum_{n=-\infty}^{+\infty} g_n e^{i(n-m)\omega t} \right\} dt \tag{A.2}$$

but
$$\oint e^{i(n-m)\omega t} \, dt = 0 \text{ or } \tau \text{ (for } m \neq n \text{ or } m = n) \tag{A.3}$$

therefore, as required
$$\oint f(t) e^{-im\omega t} \, dt = g_m \tau. \tag{A.4}$$

For the sine and cosine form of expansion we have analogously four blocks of equations as below.

If
$$f(t) = a_0 + \sum_{n=1}^{\infty} \{ a_n \cos n\omega t + b_n \sin n\omega t \} \tag{A.5}$$

215

then

$$\oint f(t)\,dt = \oint [a_0 + \Sigma\,\{a_n \cos n\omega t + b_n \sin n\omega t\,\}]\,dt$$
$$\oint f(t)\cos m\omega t\,dt = \oint \cos m\omega t\,[a_0 + \Sigma\,\{a_n \cos n\omega t + b_n \sin n\omega t\}]\,dt \quad (A.6)$$
$$\oint f(t)\sin m\omega t\,dt = \oint \sin m\omega t\,[a_0 + \Sigma\,\{a_n \cos n\omega t + b_n \sin n\omega t\}]\,dt$$

but

$$\oint \cos m\omega t\,dt = \oint \sin m\omega t\,dt = \oint \sin n\omega t \cos m\omega t\,dt = 0$$
$$\oint \cos n\omega t \cos m\omega t\,dt = \oint \sin n\omega t \cos m\omega t\,dt = \begin{cases} 0\ (m \neq n) \\ \tau/2\ (m = n) \end{cases} \quad (A.7)$$

therefore, as required,

$$\oint f(t)\,dt = a_0\tau$$
$$\oint f(t)\cos m\omega t\,dt = a_m\tau/2 \quad (A.8)$$
$$\oint f(t)\sin m\omega t\,dt = b_m\tau/2.$$

Appendix B—Delta Functions

The delta function, $\delta(x)$, may usefully, but rather imprecisely, be thought of as a function equal to zero unless $x = 0$, in which case it equals infinity, the area under this infinite spike being unity. The validity of an equation involving a delta function may be tested by considering the delta function as the limiting case of an ordinary function. A few such useful representations are:

$$\delta(x) = \lim_{a \to \infty} \sqrt{\frac{a}{\pi}}\,e^{-ax^2} \quad (B.1)$$

$$= \lim_{a \to \infty} \frac{a}{2}\,e^{-a|x|} \quad (B.2)$$

$$= \lim_{a \to 0} \frac{a}{\pi(a^2 + x^2)} \quad (B.3)$$

$$= \lim_{a \to \infty} \frac{\sin ax}{\pi x} \quad (B.4)$$

$$= \lim_{a \to \infty} \frac{1}{2\pi} \int_{-a}^{+a} e^{\pm ixy}\,dy \quad (B.5)$$

$$= \lim_{a \to \infty} \frac{\sin^2 ax}{\pi ax^2} \tag{B.6}$$

$$= \lim_{a \to \infty} (1 \mp i) \sqrt{\frac{a}{2\pi}} e^{\pm iax^2}. \tag{B.7}$$

(B.4) needs comment since although the value of the function tends to infinity at $x = 0$ as $a \to \infty$, and its area remains constant equal to unity it is not true that its value tends to zero for x not zero. However it oscillates increasingly fast with x as $a \to \infty$ and this is a sufficient condition to make it a satisfactory representation in most contexts; it means that its average value over any finite range of x not including the origin tends to zero. (B.5) integrates to give (B.4). (B.7) is also difficult to interpret as a spike of unit area; however it does satisfy (B.8) below. Other ways of seeing that (B.7) might plausibly represent a delta function are to note that its Fourier Transform tends to a constant (see Chapter 2 eqn (2.88)) or, rather doubtfully, to replace a in (B.1) by $\pm ia$.

Using $\delta(x - x_0)$ to represent a 'spike' of unit area at $x = x_0$ the following are easily verified, and (B.8) represents a rigorous and generally accepted way of defining a delta function.

$$\int f(x) \delta(x) \, dx = f(0) \tag{B.8}$$

$$\int f(x) \delta(x - x_0) \, dx = f(x_0) \tag{B.9}$$

$$\delta([x/a] - x_0) = a \, \delta(x - ax_0). \tag{B.10}$$

The integrations in (B.8) and (B.9) are over any finite or infinite region of x which encloses the position of the spike. Note that the necessity for the factor a in front of the right hand side of (B.10) is not immediately obvious, but its correctness may be confirmed by using one or other of the representations (B.1)–(B.7).

As a special application of (B.5) and (B.9) we have the commonly useful result

$$\int_{-\infty}^{+\infty} \int_{-\infty}^{+\infty} f(y) e^{\pm ix(y - y_0)} \, dx \, dy$$

$$= 2\pi \int_{-\infty}^{+\infty} f(y)\delta(y - y_0) dy = 2\pi f(y_0). \tag{B.11}$$

An infinite set of delta functions at positions $x_n = nx_0$, $(n = 0, \pm 1, \pm 2, \ldots)$ may be represented by the series:

$$\sum_{n=-\infty}^{+\infty} \delta(x - nx_0) = \frac{1}{x_0} \sum_{m=-\infty}^{+\infty} e^{\pm 2\pi imx/x_0} \qquad (B.12)$$

or equivalently

$$\sum_{m=-\infty}^{+\infty} e^{\pm imxa} = \frac{2\pi}{a} \sum_{n=-\infty}^{+\infty} \delta\left(x - n\frac{2\pi}{a}\right). \qquad (B.13)$$

These useful relations may be derived by working out the Fourier series representation of the periodic function $\delta(x - nx_0)$; see eqn (1.25).

The extension of the delta function concept to two or more dimensions is natural and is discussed in Sections 3.4 and 3.5. For instance a point delta function at the origin of a three-dimensional system may be expressed in cartesian co-ordinates, starting from one-dimensional expressions, by means of the product $\delta(\mathbf{r}) = \delta(x, y, z) = \delta(x)\delta(y)\delta(z)$. The factor 2π on the R.H.S. of eqn (B.11) has to be replaced by $(2\pi)^n$ if x and y are replaced by n dimensional vectors.

The delta function is of course not an ordinary function, but belongs to a class of functions known as generalised functions (functionals) or distributions. Its correct use and rigorous treatment in some situations requires considerable care. Nevertheless one beauty of the function lies precisely in the fact that a rather intuitive handling so often gives the correct answer.

Appendix C—The Fourier Integral Inversion Theorem

(i) *Fourier integral inversion derived as limiting case of Fourier series expansion.*

For a periodic function $f(t)$ of period τ we have shown in Appendix A that

if
$$f(t) = \sum_{n=-\infty}^{+\infty} g_n e^{+in\omega_0 t} \qquad [\omega_0 = 2\pi/\tau] \qquad (C.1)$$

then
$$g_n = \frac{1}{\tau} \int_{-\tau/2}^{+\tau/2} f(t) e^{-in\omega_0 t}. \qquad (C.2)$$

Indeed eqn (C.1) will hold for any function (not necessarily periodic) within the range $\tau/2 < t < +\tau/2$. By letting τ tend to infinity we attempt a generalisation which will suit any function. In the limit as $\tau \to \infty$ (i.e.

$\omega_0 \to 0$, $g_n \to 0$) let us assume that the continuous variable ω and the function $E(\omega)$ may be introduced as below:

as
$$\tau \to \infty \tag{C.3}$$

so
$$n\omega_0 \to \omega \tag{C.4}$$

and
$$g_n\tau \to F(\omega). \tag{C.5}$$

From (A.4) we may now write

$$F(\omega) = \lim_{\tau \to \infty} \int_{-\tau/2}^{+\tau/2} f(t)\, e^{-in\omega_0 t}\, dt \tag{C.6}$$

$$= \int_{-\infty}^{+\infty} f(t)\, e^{-i\omega t}\, dt. \tag{C.7}$$

Similarly (A.1) may be written

$$f(t) = \lim_{\omega_0 \to 0} \sum_{n=-\infty}^{+\infty} g_n\, e^{+in\omega_0 t} = \frac{1}{2\pi} \int_{-\infty}^{+\infty} F(\omega)\, e^{i\omega t}\, d\omega. \tag{C.8}$$

Equations (C.7) and (C.8) represent the required pair of relations.

(ii) *Fourier integral derived by the use of the delta function.*

If
$$f(t) = \int_{-\infty}^{+\infty} E(\omega)\, e^{i\omega t}\, d\omega \tag{C.9}$$

then
$$\int_{-\infty}^{+\infty} f(t)\, e^{-i\omega'_0 t}\, dt = \int_{-\infty}^{+\infty} \int_{-\infty}^{+\infty} E(\omega)\, e^{i(\omega - \omega'_0)t}\, dt\, d\omega. \tag{C.10}$$

$$= 2\pi \int_{-\infty}^{+\infty} E(\omega)\delta(\omega - \omega'_0)\, d\omega \tag{C.11}$$

$$= 2\pi E(\omega'_0) \tag{C.12}$$

or
$$E(\omega) = \frac{1}{2\pi} \int_{-\infty}^{+\infty} f(t)\, e^{-i\omega t}\, dt. \tag{C.13}$$

Equations (A.9) and (A.13) constitute the required pair of relations. Equation (C.11) follows from eqn (C.10) making use of the delta function expression (B.5) and then eqn (C.12) follows making use of eqn (B.9).

Appendix D—Expressions for $FT^+ FT^+ \{f(x)\}$ and $FT^- FT^- \{f(x)\}$

$$FT^+FT^+\{f(x)\} = \frac{1}{(2\pi)^2} \int e^{+ixy} \left\{ \int e^{+ixy} f(x)\, dx \right\} dy \qquad (D.1)$$

$$= \frac{1}{(2\pi)^2} \int \int e^{+iy(x+x_1)} f(x_1)\, dx_1\, dy \qquad (D.2)$$

$$= \frac{1}{(2\pi)^2} \{2\pi f(-x)\} \qquad (D.3)$$

$$= \frac{1}{2\pi} f(-x). \qquad (D.4)$$

Equation (D.3) follows from (D.2) if use is made of the delta function expressions (B.5) or (B.11) Equation (D.4) is the required result. An analogous procedure to that just used establishes also that

$$FT^-FT^-\{f(x)\} = 2\pi f(-x). \qquad (D.5)$$

An alternative procedure is to establish first one of the following relations

$$FT^+ f(x) = \frac{1}{2\pi} FT^- f(-x) \qquad (D.6)$$

$$FT^- f(x) = 2\pi\, FT^+ f(-x) \qquad (D.7)$$

and then to derive eqns (D.4) and (D.5) by multiplying eqns (D.6) and (D.7) respectively by FT^+ and by FT^-, making use of the fact that $FT^+FT^- = 1$.

Equation (D.6) may be established for instance as follows, using a dummy variable X.

$$FT^+ f(x) = \frac{1}{2\pi} \int_{-\infty}^{+\infty} e^{ixy} f(x)\, dx$$

$$= \frac{1}{2\pi} \int_{-\infty}^{+\infty} e^{iXy} f(X)\, dX$$

$$= -\frac{1}{2\pi} \int_{+\infty}^{-\infty} e^{-ixy} f(-x)\, dx \qquad [X = -x]$$

$$= \frac{1}{2\pi} \int_{-\infty}^{+\infty} e^{-ixy} f(-x)\, dx = \frac{1}{2\pi} FT^- f(-x). \qquad (D.8)$$

Appendix E—Parseval's Theorem

This theorem states that fro Fourier pairs $f(x) \leftrightarrow F(y)$ and $S(x) \leftrightarrow G(y)$,

$$\int_{-\infty}^{+\infty} f(x)g^*(x)\,dx = \frac{1}{2\pi} \int_{-\infty}^{+\infty} F(y)G^*(y)\,dy. \tag{E.1}$$

Proof

$$f(x)g^*(x) = \left\{ \frac{1}{2\pi} \int F(y_1)\,e^{+ixy_1}\,dy_1 \right\} \left\{ \frac{1}{2\pi} \int G^*(y_2)\,e^{-ixy_2}\,dy_2 \right\} \tag{E.2}$$

$$= \frac{1}{(2\pi)^2} \iint F(y_1)G^*(y_2)\,e^{-ix(y_2-y_1)}\,dy_1\,dy_2 \tag{E.3}$$

$$= \frac{1}{(2\pi)^2} \iint F(y_1)G^*(y_1+y)\,e^{-ixy}\,dy\,dy_1 \quad [y = y_2 - y_1] \tag{E.4}$$

$$\int f(x)g^*(x)\,dx = \frac{1}{(2\pi)^2} \iiint F(y_1)G^*(y_1+y)\,e^{-ixy}\,dy\,dy_1\,dx \tag{E.5}$$

$$= \frac{1}{2\pi} \int F(y_1)G^*(y_1)\,dy_1. \tag{E.6}$$

(E.3) follows from (E.2) analogously with the result that the product of two summations becomes a double summation. For instance $(\Sigma x_n)(\Sigma y_m) = \Sigma\Sigma x_n y_m$. The transition to a product of integrals may be achieved by letting the terms in the summations get progressively closer. (E.4) follows from (E.3) by the charge of variable $y = y_2 - y_1$. (E.6) follows from (E.5) using the result (B.11).

The result also holds in multi-dimensional transforms, giving for instance in three dimensions,

$$\iiint f(\mathbf{r})g^*(\mathbf{r})\,d\mathbf{r} = \frac{1}{(2\pi)^3} \iiint F(\mathbf{k})G^*(\mathbf{k})\,d\mathbf{k}. \tag{E.7}$$

A straightforward extension of Parseval's theorem shows that

$$\int_{-\infty}^{+\infty} \left(\frac{d^n\{f(x)\}}{dx^n} \right) \left(\frac{d^n\{g(x)\}}{dx^n} \right)^* dx = \frac{1}{2\pi} \int_{-\infty}^{+\infty} y^{2n} F(y)G^*(y)\,dy. \tag{E.8}$$

We prove this as follows:

$$\int \left(\frac{d^n\{f(x)\}}{dx^n}\right)\left(\frac{d^n\{g(x)\}}{dx^n}\right)^* dx$$

$$= \frac{1}{2\pi}\int \left[FT^-\left\{\frac{d^nf(x)}{dx^n}\right\}\right]\left[FT^-\frac{d^ng(x)}{dx^n}\right]^* dy \qquad (E.9)$$

$$= \frac{1}{2\pi}\int [(iy)^nF(y)][(iy)^nG(y)]^* dy \qquad (E.10)$$

$$= \frac{1}{2\pi}\int y^{2n}F(y)G^*(y)\, dy. \qquad (E.11)$$

(E.9) is a statement of Parseval's theorem and (E.10) follows using eqn (2.29).

Appendix F—The Schwartz Inequality and the Bandwidth Theorem

The bandwidth theorem, relating the widths of functions $f(x)$ and $F(y)$ as in eqn (2.28), may be proved making use of Parseval's theorem (Appendix E) and of the Schwartz inequality. The Schwartz inequality for two functions $f(x)$ and $g(x)$ can be stated in the two alternative but equivalent forms

$$\int |f|^2\, dx \int |g|^2\, dx \geqslant |\int fg^*\, dx|^2 \qquad (F.1)$$

$$\geqslant |\int fg\, dx|^2. \qquad (F.2)$$

This may be derived from the obvious inequality in eqn (F.3), the integrations throughout being understood to cover the same limits from $x = a$ to $x = b$ where a and b are arbitrary.

$$0 \leqslant \int |f - Ag|^2\, dx \qquad (F.3)$$

$$\leqslant \int |f|^2\, dx + |A|^2 \int |g|^2\, dx - A^* \int fg^*\, dx - A \int f^*g\, dx. \qquad (F.4)$$

If now the arbitrary constant A is put equal to $\int |f|^2\, dx / \int f^*g\, dx$ then on substituting this value into eqn (F.4) and dividing throughout by $\int |f|^2\, dx$ we obtain

$$0 \leqslant 1 + \frac{\int |f|^2\, dx \int |g|^2\, dx}{|\int fg^*\, dx|^2} - 1 - 1 \qquad (F.5)$$

and eqn (F.1) follows immediately. Equation (F.2) results simply from replacing $g(x)$ by its complex conjugate throughout. It is easily verified that the equalities eqns (F.1) and (F.2) will hold respectively if

$$f(x) = Kg(x) \tag{F.6}$$

or

$$f(x) = Kg^*(x) \tag{F.7}$$

where K is any constant.

We may now seek a relationship between the second moments of a function $f(x)$ and of its Fourier transform $F(y)$. For simplicity let us take the moments about the origins in the first instance. Making use of eqns (E.11) and F.1) in turn we may proceed as follows, the integrations being understood to run from $-\infty$ to $+\infty$ in all cases.

$$\int x^2 |f|^2 \, dx \int y^2 |F|^2 \, dy = 2\pi \int x^2 |f|^2 \, dx \int \left| \frac{df}{dx} \right|^2 \, dx \tag{F.8}$$

$$\geqslant 2\pi \left| \int x f \frac{df^*}{dx} \, dx \right|^2 . \tag{F.9}$$

If for convenience we write this last integral as

$$\int x f \frac{df^*}{dx} \, dx = I \tag{F.10}$$

then integration by parts establishes that provided $x|f|^2 \to 0$ as $x \to \infty$ then

$$I = -I^* - \int |f|^2 \, dx. \tag{F.11}$$

Since the modulus of a complex number is greater than or equal to the magnitude of its real part we may write, using eqn (F.11) and then Parseval's theorem (eqn (2.27)),

$$|I|^2 \geqslant \left| \frac{I + I^*}{2} \right|^2 \tag{F.12}$$

$$\geqslant \frac{1}{4} \left| \int |f|^2 \, dx \right|^2 \tag{F.13}$$

$$\geqslant \frac{1}{4} \frac{1}{2\pi} \int |f|^2 \, dx \int |F|^2 \, dy. \tag{F.14}$$

Combining eqns (F.9), (F.10) and (F.14) now gives us the required result

$$\frac{\int x^2 |f|^2 \, dx}{\int |f|^2 \, dx} \frac{\int y^2 |F|^2 \, dy}{\int |F|^2 \, dy} \geqslant \frac{1}{4}. \tag{F.15}$$

It is interesting that application of the condition in eqn (F.6) to the functions used in eqn (F.8) leads to the condition

$$xf(x) = (\text{constant}) \times \frac{df}{dx} \tag{F.16}$$

from which we derive the well-known result that a Gaussian function,

$$f(x) \propto e^{-\alpha x^2}, \tag{F.17}$$

leads to an equality in eqn (F.15).

We can readily generalize the above results to the case when we take moments not about the origins but about arbitrary points x_0 and y_0, since this is equivalent to shifting the functions. If we define a Fourier pair $g(x)$ and $G(y)$ by the relations

$$g(x) = e^{-iy_0 x} f(x + x_0) \tag{F.18}$$

$$G(y) = e^{+ix_0(y + y_0)} F(y + y_0) \tag{F.19}$$

then it is easily verified using the shift theorem, eqns (2.24) and (2.25), that

$$\frac{\int (x - x_0)^2 |f(x)|^2 \, dx}{\int |f|^2 \, dx} \frac{\int (y - y_0)^2 |F(y)|^2 \, dy}{\int |F|^2 \, dy}$$

$$= \frac{\int x^2 |f(x + x_0)|^2 \, dx}{\int |f(x + x_0)|^2 \, dx} \frac{\int y^2 |F(y + y_0)|^2 \, dy}{\int |F(y + y_0)|^2 \, dy}$$

$$= \frac{\int x^2 |g(x)|^2 \, dx}{\int |g(x)|^2 \, dx} \frac{\int y^2 |G(y)|^2 \, dy}{\int |G(y)|^2 \, dy} \geqslant \frac{1}{4}. \tag{F.20}$$

This finally is the result we set out to prove.

Appendix G—The Product and Convolution Theorems

If $f(x) \leftrightarrow F(y)$ and $g(x) \leftrightarrow G(y)$ then:

$$FT^{-}\{f(x) \otimes g(x)\} = \int e^{-ixy} \left[\int f(x_1) g(x - x_1) \, dx_1 \right] dx \tag{G.1}$$

$$= \iint e^{-ix_1 y} e^{-ix_2 y} f(x_1) g(x_2) \, dx_1 \, dx_2 \quad [x - x_1 = x_2] \tag{G.2}$$

$$= F(y) G(y). \tag{G.3}$$

By an analogous set of equations we may derive likewise that

$$FT^+\{F(y) \otimes G(y)\} = 2\pi f(x)g(x). \qquad (G.4)$$

Thus

$$f(x) \otimes g(x) \leftrightarrow F(y)G(y) \qquad (G.5)$$

and

$$f(x)g(x) \leftrightarrow \frac{1}{2\pi}\{F(y) \otimes G(y)\}. \qquad (G.6)$$

The result may readily generalised to cover n-dimensional transforms in which case the factor 2π must be replaced by $(2\pi)^n$ in eqns (G.4) and (G.6).

Appendix H—Bessel Functions

The diagrams below illustrate some characteristics of the functions $J_n(x)$, Bessel functions of the first kind of order n, for n integral and x real. For further tabulations and graphs see for instance reference 8.

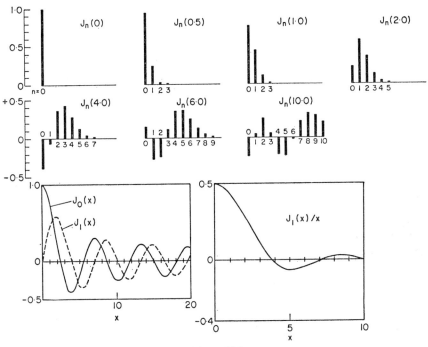

Figure H.1.

As the diagrams indicate $J_n(x)$ oscillates with increase in x (for fixed n) and at large x becomes a harmonic oscillation of decreasing amplitude. At fixed x, $J_n(x)$ has appreciable values for values of n up to an integer of about the same magnitude as x, but for values of n larger than this $J_n(x)$ becomes relatively small. $J_0(x)$ has zeroes at $x = 2 \cdot 405$, $5 \cdot 520$, $8 \cdot 654$, $11 \cdot 792$, $14 \cdot 931$, ... and $J_1(x)$ has zeroes at $x = 3 \cdot 832$, $7 \cdot 016$, $10 \cdot 173$, $13 \cdot 324$, $16 \cdot 471$, ...

The function $J_n(x)$ is important since it represents the amplitude of the nth sideband which occurs when a harmonic oscillation is phase modulated, x being proportional to the depth of the modulation and $J_0(x)$ being the amplitude of the fundamental component (see eqns (2.80)–(2.87) and (2.89)). $J_1(x)$ is important since it occurs in the two-dimensional Fourier transform of a 'circular disc', $J_1(x)/x$ being related to the angular distribution of amplitude when radiation is diffracted at a circular aperture (see eqns (3.27) and (11.78)).

Some useful relations are:

$$J_n(x) = \frac{(i)^{-n}}{2\pi} \int_0^{2\pi} \exp\left[i(x \cos \phi + n\phi)\right] \mathrm{d}\phi = \sum_{m=0}^{\infty} \frac{(-1)^m (x/2)^{n+2m}}{m!(n+m)!} \tag{H.1}$$

$$J_0(x) \approx 1 - \frac{x^2}{4} \approx \exp\left(-x^2/4\right) \qquad [x \ll 1] \tag{H.2}$$

$$J_{-n}(x) = J_n(-x) = (-1)^n J_n(x) \tag{H.3}$$

$$\exp\left(ia \sin bx\right) = \sum_{n=-\infty}^{+\infty} J_n(a)\,\mathrm{e}^{inbx} \tag{H.4}$$

$$\sum_{n=-\infty}^{+\infty} J_n(x) = \sum_{n=-\infty}^{+\infty} |J_n(x)|^2 = 1 \tag{H.5}$$

$$\int_0^X x J_0(x)\,\mathrm{d}x = X J_1(X). \tag{H.6}$$

Bessel functions are not restricted to integral order and real argument. Indeed the function $J_\nu(z)$ is defined, for ν and z complex, as one solution of Bessel's equation:

$$z^2 \frac{\mathrm{d}^2 y}{\mathrm{d}z^2} + z \frac{\mathrm{d}y}{\mathrm{d}z} + (z^2 - \nu^2)y = 0. \tag{H.7}$$

A complete solution can be written as

$$y = A J_\nu(z) + B Y_\nu(z) \tag{H.8}$$

where the functions $Y_\nu(z)$ are known as Bessel functions of the second kind of order ν and argument z. For a discussion of these more general aspects of Bessel functions see for instance reference 22. Some results useful for the applications discussed in this book are as follows:

$$J_{\frac{1}{2}}(x) = \sqrt{\frac{2}{\pi x}} \sin x \qquad \text{(H.9)}$$

$$J_{-\frac{1}{2}}(x) = \sqrt{\frac{2}{\pi x}} \cos x \qquad \text{(H.10)}$$

$$J_n(ix) = (i)^n I_n(x) \qquad \text{(H.11)}$$

where in eqn (H.11) we take n to be integral and x real, and where $I_n(x)$ is a real function known as the hyperbolic Bessel function. $I_n(x)$ is tabulated for instance in reference 8.

Appendix I—Transforms of Symmetrical Functions in Two and Three Dimensions

1. *Two dimensions*
   If θ is the angle between $\mathbf{k}$ and $\mathbf{r}$ as in Fig. I.1, then

$$d\mathbf{r} = r \, d\theta \, dr$$

$$\mathbf{k} \cdot \mathbf{r} = kr \cos \theta$$

so that if $f(\mathbf{r}) = f(r)$ then

$$\int \exp(i\mathbf{k} \cdot \mathbf{r}) f(\mathbf{r}) \, d\mathbf{r} = \int_0^\infty dr \int_0^{2\pi} d\theta \, \{\exp(ikr \cos \theta) \, r f(r)\} \qquad \text{(I.1)}$$

$$= \int_0^\infty 2\pi r f(r) J_0(kr) \, dr \qquad \text{(I.2)}$$

as required. Equation (I.2) follows from the standard result that

$$\int_0^{2\pi} \exp(ix \cos \theta) \, d\theta = 2\pi J_0(x). \qquad \text{(I.3)}$$

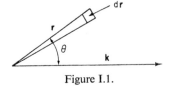

Figure I.1.

2. Three dimensions

If θ and ϕ are based on spherical polar co-ordinates as in Fig. I.2 then

$$d\mathbf{r} = r^2 \sin \theta \, dr \, d\theta \, d\phi$$

$$\mathbf{k} \cdot \mathbf{r} = kr \cos \theta$$

so that if $f(\mathbf{r}) = f(r)$ then

$$\int \exp (i\mathbf{k} \cdot \mathbf{r}) f(\mathbf{r}) \, d\mathbf{r}$$

$$= \int_0^\infty dr \int_0^\pi d\theta \int_0^{2\pi} d\phi \, \{\exp (ikr \cos \theta) f(r) r^2 \sin \theta\} \tag{I.4}$$

$$= \int_0^\infty dr \int_{-1}^{+1} dx \, [2\pi r^2 f(r)\{\cos (krx) + i \sin (krx)\}] \qquad [x = \cos \theta] \tag{I.5}$$

$$= \int_0^\infty \left\{4\pi r^2 f(r) \frac{\sin kr}{kr}\right\} dr \tag{I.6}$$

as required.

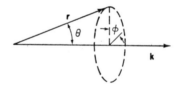

Figure I.2.

Appendix J—The Wiener–Khintchine Theorem

The following proof of the theorem is closely analogous to that used in Appendix G to prove the convolution theorem.

If $f(x) \leftrightarrow F(y)$ and $g(x) \leftrightarrow G(y)$ then:

$$FT^-\{f(x) * g(x)\}$$

$$= \int e^{-ixy} \left[\int \int f^*(x_1) g(x_1 + x) \, dx_1 \right] dx \tag{J.1}$$

$$= \int e^{+ix_1 y} e^{-ix_2 y} f^*(x_1) g(x_2) \, dx_1 \, dx_2 \qquad [x_1 + x = x_2] \tag{J.2}$$

$$= F^*(y)G(y). \tag{J.3}$$

Using an analogous set of equations we may derive likewise that

$$FT^+\{F(y) * G(y)\} = 2\pi f^*(x) g(x). \tag{J.4}$$

If we define the cross-energy spectrum of $f(x)$ and $g(x)$ by

$$S_{f_g}(y) = F^*(y)G(y) \tag{J.5}$$

then eqn (J.3) may be written

$$\rho_{f_g}(x) \leftrightarrow S_{f_g}(y) \tag{J.6}$$

which for the special case $f(x) = g(x)$ reduces to

$$\rho_f(x) \leftrightarrow S_f(y). \tag{J.7}$$

Equations (J.6) and (J.7) constitute the Wiener–Khintchine theorem.

The extension of the above proof to n-dimensional transforms is readily performed if the multi-dimensional integrals are written in their cartesian forms and separated into products of several one-dimensional integrals. As a result the factor 2π in eqn (J.4) becomes replaced by $(2\pi)^n$.

If $f(x)$ and $g(x)$ are finite power functions then the quantities in eqns (J.6) and (J.7) become infinite and we have to deal with power spectra instead of energy spectra. If $f_X(x)$ and $g_X(x)$ are truncated versions of $f(x)$ and $g(x)$, equal to $f(x)$ and $g(x)$ respectively for $|x| \leqslant X$ but equal to zero otherwise, and if $F_X(y)$ and $G_X(y)$ are the corresponding Fourier transforms so that

$$f_X(x) \leftrightarrow F_X(y) \tag{J.8}$$

$$g_X(x) \leftrightarrow G_X(y) \tag{J.9}$$

then from eqn (J.3):

$$f_X(x) * g_X(x) \leftrightarrow F_X^*(y)G_X(y) \tag{J.10}$$

$$\lim_{X \to \infty} \frac{1}{2X} \{f_X(x) * g_X(x)\} \leftrightarrow \lim_{X \to \infty} \frac{1}{2X} \{F_X^*(y)G_X(y)\} \tag{J.11}$$

$$R_{f_g}(x) \leftrightarrow P_{f_g}(y) \tag{J.12}$$

where the quantity $P_{f_g}(y)$ is called the cross-power spectrum of $f(x)$ and $g(x)$ and is defined as equal to the limit on the right hand side of eqn (J.11). If $f(x) = g(x)$ the following special case results:

$$R_f(x) \leftrightarrow P_f(y). \tag{J.13}$$

Appendix K—The Autocorrelation Function of a Stationary Random Function

If we define $\langle \Delta_f^2(t) \rangle$ as the mean square change in value over time t of a a real stationary random function $f(t)$, then

$$\langle \Delta_f^2(t) \rangle = \lim_{T \to \infty} \frac{1}{2T} \int_{-T}^{+T} \{f(t_1 + t) - f(t_1)\}^2 \, dt_1 \tag{K.1}$$

$$= \lim_{T \to \infty} \frac{1}{2T} \int_{-T}^{+T} \{f^2(t_1 + t) + f^2(t_1) - 2f(t_1)f(t_1 + t)\} \, dt_1 \tag{K.2}$$

$$= 2\langle f^2(t) \rangle - 2R_f(t) \tag{K.3}$$

and

$$R_f(t) = \langle f^2(t) \rangle - \tfrac{1}{2}\langle \Delta_f^2(t) \rangle. \tag{K.4}$$

For sufficiently short time intervals we may write:

$$\langle \Delta_f^2(t) \rangle = \langle \{f(t_1 + t) - f(t)\}^2 \rangle \tag{K.5}$$

$$\approx \langle \{tf'(t_1)\}^2 \rangle \tag{K.6}$$

where $f'(t)$ is the time differential of $f(t)$. Thus for sufficiently small times we may approximate (K.4) to give

$$R_f(t) \approx \langle f^2(t) \rangle - \tfrac{1}{2}t^2 \langle \{f'(t)\}^2 \rangle. \tag{K.7}$$

Appendix L—Spectrum and Autocorrelation Function For Shot Noise

Before proceeding to the signal $f(t)$ of infinite duration considered in Section 6.5 (eqn (6.20)) let us consider a finite section of such a signal extending from $t = 0$ to $t = T$ calling this function $f_T(t)$. Let the period T be long enough to contain a large number N of events, and let it also be large compared with the duration of the basic signal $h(t)$. Letting $F(\omega)$, $F_T(\omega)$ and $H(\omega)$ be the transforms of $f(t)$, $f_T(t)$ and $h(t)$ we get

$$f_T(t) = \sum_{i=1}^{N} h(t - t_i) \tag{L.1}$$

$$F_T(\omega) = H(\omega) \sum_{i=1}^{N} e^{-i\omega t_i} \tag{L.2}$$

$$F_T(\omega)\,F_T{}^*(\omega) = H(\omega)\,H^*(\omega) \sum_{i=1}^{N} \sum_{j=1}^{N} \exp\left[-\,i\omega(t_i - t_j)\right] \qquad (L.3)$$

$$= H(\omega)\,H^*(\omega) \left\{ N + \sum_{i \neq j} \exp\left[-\,i\omega(t_i - t_j)\right] \right\}. \qquad (L.4)$$

For N large and $|\omega| \gg 1/T$ the double summation in eqn (L.4) will represent a complex number with amplitude of order $\sqrt{N}$ if the t_i and t_j are randomly spread. This is a standard result in probability theory, one useful way of seeing it being to notice that on an Argand diagram the summation is equivalent to a random walk of N steps of unit length. The root mean square distance travelled in such a walk is well known to be equal to $\sqrt{N}$. On letting N and T tend to infinity in eqn (L.4) we may derive the following, valid for non zero ω,

$$P_f(\omega) = \lim_{T \to \infty} \frac{1}{T} F_T(\omega)\,F_T{}^*(\omega) = \lim_{T \to \infty} \frac{N}{T} H(\omega)\,H^*(\omega)$$

$$= \nu S_h(\omega) \qquad [\omega \neq 0]. \qquad (L.5)$$

This result will be generally true for all ω except for the possibility of a delta function at $\omega = 0$. Just such a singularity does occur if the mean value of $f(t)$ is non-zero as discussed in Section 6.3; using eqn (6.17) we may now generalize eqn (L.5) to give

$$P_f(\omega) = \nu S_h(\omega) + 2\pi\,|\langle f(t)\rangle|^2\,\delta(\omega) \qquad (L.6)$$

so that on taking the inverse Fourier transform we get:

$$R_f(t) = \nu\rho_h(t) + |\langle f(t)\rangle|^2. \qquad (L.7)$$

If the events are correlated so that for an event at time t_1 the probability of another event between time $t_1 + t$ and $t_1 + t + \mathrm{d}t$ is, say, $P(t)\,\mathrm{d}t$ (and correspondingly for the truncated sample we have a probability $P_T(t)\,\mathrm{d}t$ such that $\int_{-\infty}^{+\infty} P_T(t)\,\mathrm{d}t = N - 1$) then eqn (L.4) may be written as:

$$F_T(\omega)F_T{}^*(\omega) = H(\omega)\,H^*(\omega) \left\{ N + N \int_{-\infty}^{+\infty} \mathrm{e}^{-i\omega t} P_T(t)\,\mathrm{d}t \right\}. \qquad (L.8)$$

After taking the limit as $T \to \infty$ we obtain as the analogue of (L.5)

$$P_f(\omega) = \nu S_h(\omega) \left\{ 1 + \int_{-\infty}^{+\infty} \mathrm{e}^{-i\omega t} P(t)\,\mathrm{d}t \right\} \qquad (L.9)$$

which on Fourier inversion gives

$$R_f(t) = \nu\rho_h(t) + \nu\rho_h(t) \otimes P(t). \qquad (L.10)$$

These results are analogues to those discussed in Section (13.6) where diffraction from an array of point scatterers whose positions are correlated is discussed.

Appendix M—Random Phase Modulation

(i) *General Formula* (*see Section* 6.7, *eqn* (6.45)).

$$R_f(t) = \lim_{T \to \infty} \frac{1}{2T} \int_{-T}^{+T} \exp\left[-i\left\{\omega_0 t_1 + \phi(t_1)\right\}\right]$$

$$\times \exp\left[+i\left\{\omega_0(t_1 + t) + \phi(t_1 + t)\right\}\right] dt_1 \qquad (\text{M.1})$$

$$= \lim_{T \to \infty} \frac{1}{2T} \int_{-T}^{+T} \exp(i\omega_0 t) \exp\left[i\left\{\phi(t_1 + t) - \phi(t_1)\right\}\right] dt_1 \qquad (\text{M.2})$$

$$= e^{i\omega_0 t} \lim_{T \to \infty} \frac{1}{2T} \int_{-T}^{+T} dt_1 \int_{-\infty}^{+\infty} d\phi \{e^{i\phi} P(\phi, t_1, t)\} \qquad (\text{M.3})$$

$$= e^{i\omega_0 t} \int_{-\infty}^{+\infty} e^{i\phi} P(\phi, t) \, d\phi. \qquad (\text{M.4})$$

Equations (M.1) and (M.2) follow from the definition of autocorrelation function. In eqn (M.3) the function $P(\phi, t_1, t) \, d\phi$ represents the probability that in a time interval t starting at time t_1 the phase, $\phi(t)$, will have changed by an amount in the range ϕ to $\phi + d\phi$. The step from eqn (M.2) to (M.3) is not immediately obvious, but the reader may convince himself that they contain alternative ways of expressing the following mean value (averaged over t_1):

$$\langle \exp\left[i\left\{\phi(t_1) - \phi(t_1 + t)\right\}\right]\rangle.$$

Since $P(\phi, t_1, t)$ will in fact be independent of t_1 in a stationary random process we may abbreviate this to $P(\phi, t)$ and arrive at eqn (M.4), as required.

(ii) *Weak modulation* (*see Section* 6.7, *eqns* (6.51) *and* (6.52)).

Starting from eqn (M.2) above we may, if $\phi(t) \ll 1$, expand the exponential to obtain:

$$R_f(t) = e^{i\omega_0 t} \lim_{T \to \infty} \frac{1}{2T} \int_{-T}^{+T} \left[1 + i\left\{\phi(t_1 + t) - \phi(t_1)\right\}\right. \qquad (\text{M.5})$$

$$\left. - \tfrac{1}{2}\left\{\phi^2(t_1 + t) + \phi^2(t_1) - 2\phi(t_1)\phi(t_1 + t)\right\}\right] dt_1$$

$$= e^{i\omega_0 t} \left[1 - \langle\phi^2(t)\rangle + R_\phi(t)\right]. \qquad (\text{M.6})$$

On Fourier transforming this becomes,

$$P_f(\omega) = 2\pi\left(1 - \langle\phi^2(t)\rangle\right)\delta(\omega - \omega_0) + P_\phi(\omega - \omega_0). \qquad (\text{M.7})$$

Appendix N—Random Telegraph Signals

(i) *First type* (*see Section* 6.8).

Irrespective of the values of t_1 or t the product $f(t_1)f(t_1 + t)$ must always have the value $+A^2$ or the value $-A^2$, the former corresponding to an even number of sign changes during the interval t and the latter corresponding to an odd number. Thus on averaging:

$$\langle f(t_1)f(t_1 + t) \rangle = A^2 \times \text{(Probability of an even number of sign changes)}$$

$$- A^2 \times \text{(Probability of an odd number of sign changes).} \qquad \text{(N.1)}$$

Now the probability of n changes of sign during a period t is governed by the Poisson distribution (applicable also to radio-active counting experiments for instance) and is $(vt)^n(1/n!) \exp(-vt)$. Thus for positive t:

$$\langle f(t_1)f(t_1 + t) \rangle = A^2 e^{-vt}\left(1 + \frac{(vt)^2}{2!} + \dots\right)$$

$$- A^2 e^{-vt}\left(\frac{vt}{1!} + \frac{(vt)^3}{3!} + \dots\right) \qquad \text{(N.2)}$$

$$= A^2 e^{-2vt}. \qquad \text{(N.3)}$$

Thence we get the results

$$R_f(t) = A^2 e^{-2v|t|} \qquad \text{(N.4)}$$

$$P_f(\omega) = \frac{4A^2 v}{\omega^2 + 4v^2}. \qquad \text{(N.5)}$$

In eqn (N.4) the extension to negative values of t follows since for real functions an autocorrelation function is always an even function (as follows readily from the general definition). Equation (N.5) follows using a standard Fourier transform (eqn (2.39)).

(ii) *Second type* (*see Section* 6.9).

Irrespective of the values of t_1 or t the product $f(t_1)f(t_1 + t)$ must always equal either $+A^2$ or $-A^2$. If one (or more) reappraisals of sign has occurred during the time t then the values of $+A^2$ and $-A^2$ are equally likely giving an average value of zero. If no reappraisal occurs then the value must equal $+A^2$. For $t > \tau$ a reappraisal is necessary so that the average is zero. For $0 \leqslant t \leqslant \tau$ the chances of reappraisal of sign are t/τ and so the probability that no reappraisal has occurred is $(1 - t/\tau)$. Thus:

$$\langle f(t_1)f(t_1 + t)\rangle = 0 \qquad [t > \tau] \tag{N.6}$$

$$\langle f(t_1)f(t_1 + t)\rangle = A^2\left(1 - \frac{t}{\tau}\right) \qquad [0 \leqslant t \leqslant \tau]. \tag{N.7}$$

Bearing in mind that the autocorrelation of a real function is necessarily an even function we now have:

$$\left.\begin{aligned} R_f(t) &= A^2\left(1 - \frac{|t|}{\tau}\right) &\qquad |t| \leqslant \tau \\ &= 0 &\qquad |t| > \tau \end{aligned}\right\} \tag{N.8}$$

$$P_f(\omega) = A^2\tau \left\{\frac{\sin(\omega\tau/2)}{(\omega\tau/2)}\right\}^2. \tag{N.9}$$

(N.9) follows from (N.8) using the transform eqn (2.55).

Appendix O—The Response of a Linear System Expressed in Terms of Impulse Response

We use the nomenclature established in Section 7.2 using an arrow to point from input to output so that:

$$f(t) \to g(t)$$
$$\delta(t) \to h(t)$$
$$F(\omega) = \int e^{-i\omega t} f(t)\,dt$$
$$G(\omega) = \int e^{-i\omega t} g(t)\,dt$$
$$H(\omega) = \int e^{-i\omega t} h(t)\,dt.$$

We now derive the required result by considering a succession of impulses to be applied to a system at instants t_i and of strengths $f(t_i)$. If the number of such impulses per unit time is allowed to increase indefinitely the input becomes progressively more equivalent to an input $f(t)$.

$$\delta(t) \to h(t) \tag{O.1}$$

$$\sum_i f(t_i)\,\delta(t - t_i) \to \sum_i f(t_i)\,h(t - t_i) \tag{O.2}$$

$$\int_{-\infty}^{+\infty} f(t_1)\,\delta(t - t_1)\,dt_1 \to \int_{-\infty}^{+\infty} f(t_1)\,h(t - t_1)\,dt_1 \tag{O.3}$$

$$f(t) \to f(t) \otimes h(t) \tag{O.4}$$

so that

$$g(t) = f(t) \otimes h(t). \tag{O.5}$$

(O.2) follows from (O.1) by utilising both the linearity and shift invariance properties of the system. (O.3) follows by considering more and more delta functions in eqn (O.2) and considering the limit as the spacing tends to zero so that a sum over $f(t_i)$ is replaced by an integral over $f(t_1)$. We now utilise (B.9) to reduce the left hand side of (O.3) and note that the right-hand side defines a convolution to arrive at the required result (O.4).

Application of the convolution theorem to (O.5) gives on Fourier transformation

$$G(\omega) = F(\omega) H(\omega). \tag{O.6}$$

Appendix P—Use of Kirchhoff's Formula for Diffraction at a Plane Aperture

We consider the scalar wave equation

$$\nabla^2 \phi = \frac{1}{c^2} \frac{\partial^2 \phi}{\partial t^2} \tag{P.1}$$

and consider a monochromatic solution at frequency ω, which can be written generally in terms of functions $A(\mathbf{R})$, $\alpha(\mathbf{R})$ and $f(\mathbf{R})$ as follows

$$\phi(\mathbf{R}, t) = A(\mathbf{R}) \cos\{\omega t - \alpha(\mathbf{R})\} \tag{P.2}$$

$$= \text{Re}\,[f(\mathbf{R})\,e^{-i\omega t}] \tag{P.3}$$

where

$$f(\mathbf{R}) = A(\mathbf{R})\,e^{i\alpha(\mathbf{R})}. \tag{P.4}$$

Kirchhoff showed that the value of the complex amplitude f at any point P in a radiation field may be derived from a knowledge of f and df/dn over any closed surface S surrounding P. df/dn is here the derivative of f with respect to the outward normal to the surface, and we assume there are no obstacles or sources within the surface. Kirchhoff's rigorous expression is proved for instance in references 1, 4, 13, 17 and 18, and it may be written

$$f_P = \frac{1}{4\pi} \iint\limits_{S} \left\{ \frac{1}{r} e^{ikr} \frac{df}{dn} - f \frac{df}{dn} \left(\frac{e^{ikr}}{r} \right) \right\} dS. \tag{P.5}$$

The integral is over the surface S, and r is the distance from an element of the surface, of area dS, to the point P. k is the wave number of a free wave given by $k = \omega/c$.

We now develop an approximate version of (P.5) suitable for deriving the diffraction field from a plane diffracting aperture from a knowledge of $f(\mathbf{R})$ at points in the plane of the aperture. We choose the surface S to enclose the point P and to coincide with the plane of the aperture over the whole region of the finite sized aperture only breaking away from this plane at very large distances. It is shown in the references quoted that in integrating (P.5) over S we need only consider the portion coinciding with the aperture provided the rest of the surface is chosen to be sufficiently remote. We now write (P.5) as an integral over elements of area ds of the aperture plane in the form

$$f_P = \frac{1}{4\pi} \iint \left\{ \frac{1}{r} e^{ikr} \frac{df}{dn} - f\cos\theta \frac{d}{dr}\left(\frac{e^{ikr}}{r}\right) \right\} ds. \qquad (P.6)$$

We have taken the opportunity of transforming from d/dn to d/dr in the second term by introducing θ, the angle between the normal to the aperture plane and a line from the element ds to the point P.

We now approximate by assuming that over the plane of the aperture the amplitude f may be approximated locally to a plane wave emerging normally from the aperture, in which case

$$\frac{df}{dn} \approx -ikf \qquad (P.7)$$

and (P.6) becomes

$$f_P \approx \frac{1}{4\pi} \iint e^{ikr} \left\{ -\frac{ikf}{r} - f\cos\theta \left(\frac{ik}{r} - \frac{1}{r^2}\right) \right\} ds. \qquad (P.8)$$

If we now further approximate by assuming that $kr \gg 1$ we may obtain finally

$$f_P \approx \frac{-ik}{2\pi} \iint f\left(\frac{1 + \cos\theta}{2}\right) \frac{e^{ikr}}{r} ds. \qquad (P.9)$$

Appendix Q—Fresnel and Fraunhofer Diffraction Treated by the Angular Spectrum Method

We establish here the eqns (11.29)–(11.35) starting from eqn (11.28) and using the nomenclature established in Section (11.2).

From equation (11.28) we have

$$F(\mathbf{q}, z) = \exp\left(iz\sqrt{k^2 - q^2}\right) F(\mathbf{q}, 0) \qquad (Q.1)$$

so that on Fourier inversion we get (see eqn (3.22))

$$f(\mathbf{S}, \mathbf{z}) = f(\mathbf{S}, 0) \otimes \left[\frac{1}{(2\pi)^2} \iint \exp{(i\mathbf{q} \cdot \mathbf{S})} \{\exp{(iz\sqrt{k^2 - q^2}\}} \, d\mathbf{q} \right]. \quad (Q.2)$$

We may expand the index in eqn (Q.1) as a power series in ascending powers of q/k:

$$z\sqrt{k^2 - q^2} = zk - \frac{zq^2}{2k} - \frac{zq^4}{zk^3} + \dots \quad (Q.3)$$

If now $F(\mathbf{q}, 0)$ is a narrow angle distribution so that we may assume $F(\mathbf{q}, 0) = 0$ for $q > q_M$, where q_M is some maximum value, then for

$$\frac{zq_M^4}{8k^3} \ll 1 \quad \text{[Fresnel region]} \quad (Q.4)$$

we may use the first two terms only of eqn (Q.3) and write eqn (Q.1) as

$$F(\mathbf{q}, z) = e^{ikz} \{e^{-izq^2/2k} F(\mathbf{q}, 0)\}. \quad (Q.5)$$

Using eqn (3.25a) together with the product formula, eqn (3.22), we obtain on Fourier inverting equation (Q.5):

$$f(\mathbf{S}, z) = e^{ikz} \left\{ f(\mathbf{S}, 0) \otimes \left(-\frac{ik}{2\pi z} \right) \exp{(ikS^2/2z)} \right\} \quad (Q.6)$$

$$= -\frac{ik}{2\pi z} e^{ikz} \iint f(\mathbf{s}, 0) \exp{[ik(\mathbf{S} - \mathbf{s})^2/2z]} \, d\mathbf{s} \quad (Q.7)$$

$$= -\frac{ik}{2\pi z} \exp{[i(kz + kS^2/2z)]} \iint \exp{(-ik\mathbf{S} \cdot \mathbf{s}/z)}$$

$$\times \{f(\mathbf{s}, 0) \exp{(iks^2/2z)}\} \, d\mathbf{s}. \quad (Q.8)$$

We have thus derived eqns (11.30), (11.31), and (11.32) as required. Note that eqn (Q.8) is expressed as a Fourier transform with kS/z and $\mathbf{s}$ as conjugate variables.

Let us now consider a further approximation by using the expansion:

$$kr = k\sqrt{z^2 + (\mathbf{S} - \mathbf{s})^2} \quad (Q.9)$$

$$= kz + \frac{k(\mathbf{S} - \mathbf{s})^2}{2z} - \frac{k(\mathbf{S} - \mathbf{s})^4}{8z^3} + \dots \quad (Q.10)$$

If we consider an aperture function for which $f(\mathbf{s}, 0) = 0$ for $s > s_M$, and if with $s \lesssim s_M$ we consider only values of z and $\mathbf{S}$ such that

$$\frac{k(\mathbf{S} - \mathbf{s})^4}{8z^3} \ll 1 \qquad (Q.11)$$

then we may ignore the last term in eqn (Q.10) and rewrite eqn (Q.7) as

$$f(\mathbf{S}, z) = -\frac{ik}{2\pi z} \iint f(\mathbf{s}, 0)\, e^{ikr}\, d\mathbf{s}. \qquad (Q.12)$$

Alternatively noting that

$$kR = k\sqrt{z^2 + S^2} \qquad (Q.13)$$

$$= kz + \frac{kS^2}{2z} - \frac{kS^4}{8z^3} + \cdots \qquad (Q.14)$$

then with the approximations

$$\frac{kS^4}{8z^3} \ll 1 \qquad (Q.15a)$$

and

$$\frac{k{s_M}^2}{2z} \ll 1 \qquad (Q.15b)$$

we may rewrite eqn (Q.8) as:

$$f(\mathbf{S}, z) = -\frac{ik}{2\pi z} e^{ikR} \iint \exp\left(-ik\mathbf{S}\cdot\mathbf{s}/z\right) f(\mathbf{s}, 0)\, d\mathbf{s}. \qquad (Q.16)$$

Equations (Q.12) and (Q.16) are now the required equations for the Fraunhofer region.

It is worth commenting that the approximations necessary for our equations to be valid are quite subtle it being not sufficient merely to classify the Fresnel equations as suitable for the near field, and the Fraunhofer ones for the far field. To illustrate this let us consider a case approximating to visible light of wavelength $2\pi\, 10^{-7}$m incident on a hole of diameter about 0.1 mm so that $s_M \approx 10^{-4}$m and $q_M \approx 10^{+4}$m^{-1}, $\mathbf{s}$ and $\mathbf{q}$ being conjugate variables. The condition (Q.4) leads to the requirement that $z \ll 8\,10^5$m. The condition (Q.11) leads to a further minimum requirement (for $S = 0$) that $z \gg 5\,10^{-4}$m; if off-axis points are to be considered then z must be correspondingly greater, for instance for $S \approx 2$ cm then we find $z \gg 0.6$ m. The condition (Q.15b) means that in addition to the basic requirement

$z \ll 8\,10^5$m we would require $z \gg 5$ cm due to the size of s_M, and (Q.15a) means additionally that z must have correspondingly greater values if large values of S are to be considered; actually this last restriction does not act since due to the narrow angle nature of the Fraunhofer diffraction the intensity falls to zero at given z for values of S within the limit set by (Q.15a).

Appendix R—Ghosts in Diffraction From a Distorted Array of Point Scatterers

Let $\rho_0(\mathbf{r}) = \sum_i \delta(\mathbf{r} - \mathbf{r}_i)\,[i = 1, 2, \ldots, N]$ $\qquad$ (R.1)

$$\rho'(\mathbf{r}) = \sum_i \delta(\mathbf{r} - \mathbf{r}_i') = \sum_i \delta(\mathbf{r} - \mathbf{r}_i - \mathbf{a}_0 \sin[(\mathbf{Q}\cdot\mathbf{r}_i) + \phi]) \quad \text{(R.2)}$$

$$\mathbf{r}_i' = \mathbf{r}_i + \mathbf{a}_0 \sin[(\mathbf{Q}\cdot\mathbf{r}_i) + \phi] \qquad \text{(R.3)}$$

$$F_0(\mathbf{K}) = \int \exp(-i\mathbf{K}\cdot\mathbf{r})\,\rho_0(\mathbf{r})\,d\mathbf{r} \qquad \text{(R.4)}$$

$$F'(\mathbf{K}) = \int \exp(-i\mathbf{K}\cdot\mathbf{r})\,\rho'(\mathbf{r})\,d\mathbf{r}. \qquad \text{(R.5)}$$

To relate $F'(\mathbf{K})$ to $F_0(\mathbf{K})$ proceed as follows:

$$F'(\mathbf{K}) = \int \exp(-i\mathbf{K}\cdot\mathbf{r})\,\rho'(\mathbf{r})\,d\mathbf{r}$$

$$= \int \exp(-i\mathbf{K}\cdot\mathbf{r}) \sum_i \delta(\mathbf{r} - \mathbf{r}_i - \mathbf{a}_0 \sin[(\mathbf{Q}\cdot\mathbf{r}_i) + \phi])\,d\mathbf{r} \qquad \text{(R.6)}$$

$$= \sum_i \exp\{-i\mathbf{K}\cdot(\mathbf{r}_i + \mathbf{a}_0 \sin[(\mathbf{Q}\cdot\mathbf{r}_i) + \phi])\} \qquad \text{(R.7)}$$

$$= \sum_i \left[\exp(-i\mathbf{K}\cdot\mathbf{r}_i)\left\{\sum_{n=-\infty}^{+\infty} J_n(-\mathbf{K}\cdot\mathbf{a}_0)\exp\{in[(\mathbf{Q}\cdot\mathbf{r}_i) + \phi]\}\right\}\right] \quad \text{(R.8)}$$

$$= \sum_{n=-\infty}^{+\infty} \left[J_n(-\mathbf{K}\cdot\mathbf{a}_0)\,e^{in\phi}\left\{\sum_i \exp[-i(\mathbf{K} - n\mathbf{Q})\cdot\mathbf{r}_i]\right\}\right] \qquad \text{(R.9)}$$

$$= \sum_{n=-\infty}^{+\infty} [J_n(-\mathbf{K}\cdot\mathbf{a}_0)\,e^{in\phi}\{\int \exp[-i(\mathbf{K} - n\mathbf{Q})\cdot\mathbf{r}]\,\rho_0(\mathbf{r})\,d\mathbf{r}\}] \quad \text{(R.10)}$$

$$= \sum_{n=-\infty}^{+\infty} J_n(-\mathbf{K}\cdot\mathbf{a}_0)\,e^{in\phi}\,F_0(\mathbf{K} - n\mathbf{Q}). \qquad \text{(R.11)}$$

The necessary properties of Bessel functions are summarised in Appendix H, eqn (H.4). In the above treatment the displacements are expressed relative to the *original* sites through the term $\mathbf{Q}\cdot\mathbf{r}_i$ in eqn (R.2). An interesting variant consists in relating the displacement to the *new* site. We now proceed as follows, the method being subtley different.

Let $\qquad \rho_0(\mathbf{r}) = \sum_i \delta(\mathbf{r} - \mathbf{r}_i)$ (R.12)

$$\rho''(\mathbf{r}) = \sum_i \delta(\mathbf{r} - \mathbf{r}_i'') = \sum_i \delta(\mathbf{r} - \mathbf{r}_i - \mathbf{a}_0 \sin[(\mathbf{Q} \cdot \mathbf{r}) + \phi]) \quad \text{(R.13)}$$

$$\mathbf{r}_i'' = \mathbf{r}_i + \mathbf{a}_0 \sin[(\mathbf{Q} \cdot \mathbf{r}_i'') + \phi] \tag{R.14}$$

$$F_0(\mathbf{K}) = \int \exp(-i\mathbf{K} \cdot \mathbf{r}) \rho_0(\mathbf{r}) \, d\mathbf{r} \tag{R.15}$$

$$F''(\mathbf{K}) = \int \exp(-i\mathbf{K} \cdot \mathbf{r}) \rho''(\mathbf{r}) \, d\mathbf{r}. \tag{R.16}$$

Introduce a new variable $\mathbf{r}'' = \mathbf{r} - \mathbf{a}_0 \sin[(\mathbf{Q} \cdot \mathbf{r}) + \phi]$. Then

$$\rho''(\mathbf{r}) = \rho_0(\mathbf{r}'') = \frac{1}{(2\pi)^3} \int \exp(+i\mathbf{K} \cdot \mathbf{r}'') F_0(\mathbf{K}) \, d\mathbf{K} \tag{R.17}$$

$$= \frac{1}{(2\pi)^3} \int F_0(\mathbf{K}) \exp\{+i\mathbf{K} \cdot (\mathbf{r} - \mathbf{a}_0 \sin[(\mathbf{Q} \cdot \mathbf{r}) + \phi])\} \, d\mathbf{K} \tag{R.18}$$

$$= \frac{1}{(2\pi)^3} \int \exp(+i\mathbf{K} \cdot \mathbf{r})$$

$$\times \left\{ \sum_{n=-\infty}^{+\infty} J_n(-\mathbf{K} \cdot \mathbf{a}_0) \exp\{in[(\mathbf{Q} \cdot \mathbf{r}) + \phi]\} \right\} F_0(\mathbf{K}) \, d\mathbf{K} \tag{R.19}$$

$$= \frac{1}{(2\pi)^3} \int \exp(+i\mathbf{K}' \cdot \mathbf{r})$$

$$\times \left\{ \sum_{n=-\infty}^{+\infty} J_n(-[\mathbf{K}' - n\mathbf{Q}] \cdot \mathbf{a}_0) e^{in\phi} F_0(\mathbf{K}' - n\mathbf{Q}) \right\} d\mathbf{K}' \tag{R.20}$$

$$[\mathbf{K}' = \mathbf{K} + n\mathbf{Q}].$$

Thus

$$F''(\mathbf{K}) = \sum_{n=-\infty}^{+\infty} J_n(-[\mathbf{K} - n\mathbf{Q}] \cdot \mathbf{a}_0) e^{in\phi} F_0(\mathbf{K} - n\mathbf{Q}). \tag{R.21}$$

The positions of the 'ghosts' or 'satellites' are thus the same as in the first treatment but their amplitudes are different.

Both of these treatments may be extended to cover a time dependent distortion produced by a wave travelling across the system. We replace $\rho_0(\mathbf{r})$, $\rho'(\mathbf{r})$, $\rho''(\mathbf{r})$, $F_0(\mathbf{K})$, $F'(\mathbf{K})$ and $F''(\mathbf{K})$ by $\rho_0(\mathbf{r}, t)$, $\rho'(\mathbf{r}, t)$, $\rho''(\mathbf{r}, t)$,

$F_0(\mathbf{K}, t)$, $F'(\mathbf{K}, t)$ and $F''(\mathbf{K}, t)$. We also replace ϕ throughout by $(\phi - \Omega t)$ so that eqns (R.2) and (R.13) represent travelling disturbances of wave vector $\mathbf{Q}$ and angular frequency Ω.

We now have, from eqns (R.11) and (R.21) respectively:

$$F'(\mathbf{K}, t) = \sum_{n=-\infty}^{+\infty} J_n(-\mathbf{K} \cdot \mathbf{a}_0) \exp\left[in(\phi - \Omega t)\right] F_0(\mathbf{K} - n\mathbf{Q}, t) \qquad (\text{R.22})$$

$$F''(\mathbf{K}, t) = \sum_{n=-\infty}^{+\infty} J_n(-[\mathbf{K} - n\mathbf{Q}] \cdot \mathbf{a}_0) \exp\left[in(\phi - \Omega t)\right] F_0(\mathbf{K} - n\mathbf{Q}, t). \qquad (\text{R.23})$$

If we now define functions $F_0(\mathbf{K}, \omega)$, $F'(\mathbf{K}, \omega)$ and $F''(\mathbf{K}, \omega)$ by the transforms:

$$F_0(\mathbf{K}, \omega) = \iint \exp\left[-i(\mathbf{K} \cdot \mathbf{r} - \omega t)\right] \rho_0(\mathbf{r}, t) \, d\mathbf{r} \, dt = \int e^{+i\omega t} F_0(\mathbf{K}, t) \, dt$$

$$F'(\mathbf{K}, \omega) = \iint \exp\left[-i(\mathbf{K} \cdot \mathbf{r} - \omega t)\right] \rho'(\mathbf{r}, t) \, d\mathbf{r} \, dt = \int e^{+i\omega t} F'(\mathbf{K}, t) \, dt$$

$$F''(\mathbf{K}, \omega) = \iint \exp\left[-i(\mathbf{K} \cdot \mathbf{r} - \omega t)\right] \rho''(\mathbf{r}, t) \, d\mathbf{r} \, dt = \int e^{+i\omega t} F''(\mathbf{K}, t) \, dt$$

then $F'(\mathbf{K}, \omega)$ is given by:

$$F'(\mathbf{K}, \omega) = \sum_{n=-\infty}^{+\infty} J_n(-\mathbf{K} \cdot \mathbf{a}_0) \, e^{in\phi} \int \exp\left[i(\omega - n\Omega)t\right] F_0(\mathbf{K} - n\mathbf{Q}, t) \, dt \qquad (\text{R.24})$$

$$= \sum_{n=-\infty}^{+\infty} J_n(-\mathbf{K} \cdot \mathbf{a}_0) \, e^{in\phi} F_0(\mathbf{K} - n\mathbf{Q}, \omega - n\Omega) \qquad (\text{R.25})$$

and analogously we derive

$$F''(\mathbf{K}, \omega) = \sum_{n=-\infty}^{+\infty} J_n(-[\mathbf{K} - n\mathbf{Q}] \cdot \mathbf{a}_0) \, e^{in\phi} F_0(\mathbf{K} - n\mathbf{Q}, \omega - n\Omega). \qquad (\text{R.26})$$

Appendix S—The Pair Distribution Function for Randomly Displaced Scatterers

Let $g_0(\mathbf{r})$ be the pair distribution function for a set of point scatterers initially, and let $g'(\mathbf{r})$ be the corresponding function after they have all suffered random independent displacements. Let the probability of a scatterer being displaced by a distance $\mathbf{u}$ be $p(\mathbf{u})$, by which we mean in more precise terms that the probability of a particle being displaced by a distance $\mathbf{u}$ into a volume element $d\mathbf{u}$ is $p(\mathbf{u}) \, d\mathbf{u}$.

Consider an arbitrary reference scatterer which is initially at position $\mathbf{R}$. The joint probability of there being a different particle which is at $(\mathbf{R} + \mathbf{r}_1)$

initially and which becomes displaced to $(\mathbf{R} + \mathbf{r})$ is given by the product $g_0(\mathbf{r}_1)p(\mathbf{r} - \mathbf{r}_1)$. Thus the overall probability of some scatterer, other than the reference one, ending up at $(\mathbf{R} + \mathbf{r})$ irrespective of the value of $\mathbf{r}_1$ is given by

$$\int g_0(\mathbf{r}_1)p(\mathbf{r} - \mathbf{r}_1)\,d\mathbf{r}_1 = g_0(\mathbf{r}) \otimes p(\mathbf{r}).$$

Let us call this quantity $g_1(\mathbf{r})$.

We have so far not considered the displacement of the reference scatterer itself; however this reference scatterer will suffer a random displacement and the joint probability of the reference scatterer ending up at $(\mathbf{R} + \mathbf{r}_2)$ whilst some other scatterer ends up at $(\mathbf{R} + \mathbf{r}_2 + \mathbf{r})$ will be $p(\mathbf{r}_2)g_1(\mathbf{r}_2 + \mathbf{r})$. Thus the overall probability that the arbitrary reference scatterer ends up with another scatterer a distance $\mathbf{r}$ away, irrespective of the value of $\mathbf{r}_2$ is:

$$g'(\mathbf{r}) = \int p(\mathbf{r}_2)g_1(\mathbf{r}_2 + \mathbf{r})\,d\mathbf{r}_2$$

$$= \int p(-\mathbf{r}_3)g_1(\mathbf{r} - \mathbf{r}_3)\,d\mathbf{r}_3 \qquad [\mathbf{r}_3 = -\mathbf{r}_2]$$

$$= p(-\mathbf{r}) \otimes g_1(\mathbf{r})$$

$$= p(-\mathbf{r}) \otimes p(\mathbf{r}) \otimes g_0(\mathbf{r}). \qquad (\text{S.1})$$

In deriving eqn (S.1) we have made use of the fact that the order of the functions in a convolution is immaterial (see eqn (5.3)). Since $p(\mathbf{r})$ is a real quantity it follows that

if $p(\mathbf{r}) \leftrightarrow P(\mathbf{K})$

then $p(-\mathbf{r}) \leftrightarrow P^*(\mathbf{K})$

so that on Fourier transforming eqn (S.1) we obtain as required:

$$G'(\mathbf{K}) = |P(\mathbf{K})|^2 G_0(\mathbf{K}) \qquad (\text{S.2})$$

where $g_0(\mathbf{r}) \leftrightarrow G_0(\mathbf{K})$

and $g'(\mathbf{r}) \leftrightarrow G'(\mathbf{K}).$

Appendix T—Scattering by a Time Dependent System

In Section 13.2 we have established that when an incident wave

$$\Phi_{\text{inc}}(t) = \phi_0 \cos\{\mathbf{k}_0 \cdot \mathbf{r} - \omega_0 t\} \qquad (\text{T.1})$$

is incident on a finite scatterer of density $\rho(\mathbf{r})$, then the amplitude of the scattered wave at a large distance R away may be expressed as

$$\Phi_{sc}(t) = \text{Re}\,\phi_{sc}\,e^{-i\omega_0 t} \qquad (T.2)$$

where

$$\phi_{sc} = \frac{\phi_0 a}{R}\,e^{ikR}\int \rho(\mathbf{r})\exp\left(-i\mathbf{K}\cdot\mathbf{r}\right)\mathrm{d}\mathbf{r} \qquad (T.3)$$

and a and $\mathbf{K}$ have the meanings defined in Sections 13.2 and 13.3. If the density $\rho(\mathbf{r})$ is changing on a time scale slow compared with the period of the incident radiation then we may generalize the above equations simply by writing $\rho(\mathbf{r})$ as $\rho(\mathbf{r}, t)$ and ϕ_{sc} as $\phi_{sc}(t)$. We will assume an experiment of finite duration T and will represent this by putting $\rho(\mathbf{r}, t)$ equal to zero outside such a time limit. Thus $\rho(\mathbf{r}, t)$ is square integrable in space and in time. As a result of the time dependence of ϕ_{sc} we will now find that $\Phi_{sc}(t)$ contains frequency components over a range of frequencies distributed round ω_0. If we define a function $G(\omega)$ by

$$G(\omega) = \int e^{+i\omega t}\,\phi_{sc}(t)\,\mathrm{d}t$$

$$= \frac{\phi_0 a}{R}\,e^{ikR}\int \exp\left[-i(\mathbf{K}\cdot\mathbf{r} - \omega t)\right]\rho(\mathbf{r}, t)\,\mathrm{d}\mathbf{r}\,\mathrm{d}t \qquad (T.4)$$

then on Fourier inverting this and substituting into eqn (T.2) we obtain

$$\Phi_{sc}(t) = \text{Re}\,\frac{1}{2\pi}\int \exp\left[-i(\omega_0 + \omega)t\right]G(\omega)\,\mathrm{d}\omega. \qquad (T.5)$$

This equation may be interpreted by saying that $G(\omega)$ is proportional to the amplitude of the scattered component at frequency $(\omega_0 + \omega)$, so that ω represents the increase in frequency occurring on scattering, and the energy scattered at frequency $(\omega_0 + \omega)$ will be proportional to $|G(\omega)|^2$.

Let us now seek an expression for the scattering cross-section $\sigma(\mathbf{K})$ which we define as the ratio between the total energy scattered per steradian during the whole experiment to the total incident energy per unit area. We proceed as follows, invoking Parseval's theorem (eqn (5.44)) at the last step:

$$\sigma(\mathbf{K}) = \frac{R^2\displaystyle\int_{-\infty}^{+\infty}\Phi_{sc}^2\,\mathrm{d}t}{\displaystyle\int_{-\infty}^{+\infty}\Phi_{inc}^2\,\mathrm{d}t} = \frac{\frac{1}{2}R^2\int |\phi_{sc}(t)|^2\,\mathrm{d}t}{\int \Phi_{inc}^2\,\mathrm{d}t}$$

$$= \frac{R^2 \int |\phi_{sc}(t)|^2 \, dt}{\phi_0{}^2 T}$$

$$= \frac{R^2 \int_{-\infty}^{+\infty} |G(\omega)|^2 \, d\omega}{2\pi\phi_0{}^2 T} \quad . \tag{T.6}$$

This equation now allows us to write an expression for the differential scattering cross-section $\sigma(\mathbf{K}, \omega)$ which we define as the ratio of the energy scattered for steradian per unit frequency range to the total incident energy per unit area. Thus

$$\sigma(\mathbf{K}, \omega) = \frac{R^2 |G(\omega)|^2}{2\pi\phi_0{}^2 T}$$

$$= \frac{|a|^2}{2\pi T} |\int \exp[-i(\mathbf{K} \cdot \mathbf{r} - \omega t)] \, \rho(\mathbf{r}, t) \, d\mathbf{r} \, dt|^2. \tag{T.7}$$

Appendix U—Implications of Causality

We establish here certain relations between the real and imaginary parts, $R(\omega)$ and $X(\omega)$ respectively, of a function $H(\omega)$ which arise when $H(\omega)$ is the Fourier transform of a causal function $h(t)$ for which $h(t) = 0$ when $t < 0$. If $h(t)$ possesses no singularity at $t = 0$ then:

$$X(\omega) = -\frac{1}{\pi} \int_{-\infty}^{+\infty} \frac{R(\omega_1)}{\omega - \omega_1} \, d\omega_1 \tag{U.1}$$

$$= -\frac{2}{\pi} \int_0^\infty \int_0^\infty R(\omega_1) \cos \omega_1 t \sin \omega t \, d\omega_1 \, dt \tag{U.2}$$

$$R(\omega) = \frac{1}{\pi} \int_{-\infty}^{+\infty} \frac{X(\omega_1)}{\omega - \omega_1} \, d\omega_1 \tag{U.3}$$

$$= -\frac{2}{\pi} \int_0^\infty \int_0^\infty X(\omega_1) \sin \omega_1 t \cos \omega t \, d\omega_1 \, dt. \tag{U.4}$$

Equations (U.1) and (U.3) are known as the Kramers–Kronig dispersion relations when they are applied to the scattering of radiation from atomic systems; in this context they imply a relation between attenuation in a medium and the real part of the refractive index, (see reference 4). In the context of servo-mechanisms they are often known as Bode's relations. In a mathematical context transforms of the type introduced in eqns (U.1) and (U.3) are known as Hilbert transforms.

In deriving the above equations it is convenient to introduce a pair of functions $h_e(t)$ and $h_o(t)$ which are respectively even and odd functions of time and which are closely related to $h(t)$ as follows:

$$h(t) = h_e(t) + h_o(t) \qquad \text{(U.5)}$$

so that

$$h(t) = 2h_e(t) = 2h_o(t) \qquad [t > 0). \qquad \text{(U.6)}$$

It is now easily verified that

$$h_e(t) \leftrightarrow R(\omega) \qquad \text{(U.7)}$$

$$h_o(t) \leftrightarrow iX(\omega). \qquad \text{(U.8)}$$

If we also introduce the step function $u(t)$ defined by:

$$u(t) = 1 \qquad (t \geqslant 0]$$

$$= 0 \qquad [t < 0] \qquad \text{(U.9)}$$

then we establish eqns (U.1) and (U.3) as follows, using capital letters to represent Fourier transforms as usual:

$$H(\omega) = FT^- h(t) = FT^- \{2h_e(t)u(t)\}$$

$$= \frac{1}{2\pi} \{2R(\omega) \otimes U(\omega)\} \qquad \text{[see eqn (2.34)]}$$

$$= \frac{1}{\pi} \left[R(\omega) \otimes \left\{ \pi\delta(\omega) - \frac{i}{\omega} \right\} \right] \qquad \text{[see eqn (2.77)]}$$

$$= R(\omega) - \frac{i}{\pi} \int_{-\infty}^{+\infty} \frac{R(\omega_1)}{\omega - \omega_1} \, d\omega_1. \qquad \text{(U.10)}$$

By an analogous procedure using $h_o(t)$ instead of $h_e(t)$ we may write:

$$H(\omega) = FT^- \{2h_o(t)u(t) \}$$

$$= \frac{1}{2\pi} \{2iX(\omega) \otimes U(\omega)\}$$

$$= \frac{1}{\pi} \int_{-\infty}^{+\infty} \frac{X(\omega_1)}{\omega - \omega_1} \, d\omega_1 + iX(\omega). \qquad \text{(U.11)}$$

A comparison of eqns (U.10) and (U.11) with the following

$$H(\omega) = R(\omega) + iX(\omega) \qquad \text{(U.12)}$$

leads directly to eqns (U.1) and (U.3) as required.

We establish eqns (U.2) and (U.4) by a somewhat different method as follows. If the forward and inverse transforms in eqns (U.7) and (U.8) are written out explicitly one soon obtains:

$$R(\omega) = 2\int_0^\infty h_e(t) \cos \omega t \, dt \qquad \text{(U.13)}$$

$$X(\omega) = -2\int_0^\infty h_o(t) \sin \omega t \, dt \qquad \text{(U.14)}$$

$$h_e(t) = \frac{1}{\pi}\int_0^\infty \cos \omega t \, R(\omega) \, d\omega \qquad \text{(U.15)}$$

$$h_o(t) = -\frac{1}{\pi}\int_0^\infty \sin \omega t X(\omega) \, d\omega. \qquad \text{(U.16)}$$

Noting now that for $t > 0$ $h_e(t)$ is the same as $h_o(t)$ we may combine eqns (U.13) and (U.16) and also eqns (U.14) and (U.15) to give eqns (U.4) and (U.2) as required

Note that since $R(\omega)$ and $X(\omega)$ are each determined in terms of the other it is possible to derive $h(t)$ from a knowledge of one or other function by itself, and it is readily verified that

$$h(t) = \frac{2}{\pi}\int_0^\infty R(\omega) \cos \omega t \, d\omega$$

$$= -\frac{2}{\pi}\int_0^\infty X(\omega) \sin \omega t \, d\omega \qquad [t > 0]. \qquad \text{(U.17)}$$

For a further discussion of the implications of causality see references 12 and 21.

Bibliography

1. Born, M., and Wolf, E., "Principles of Optics". Pergamon, Oxford. 1964
2. Campbell, G. A., and Foster, R. M., "Fourier Integrals for Practical Applications". Van Nostrand, New York. 1961.
3. Churchill, R. V., "Fourier Series and Boundary Value Problems". McGraw-Hill, New York. 1963.
4. Ditchburn, R. W., "Light". Blackie, London. 1963.
5. Duffin, W. J., "Electricity and Magnetism". McGraw-Hill, London. 1965.
6. Egelstaff, P. A., (editor) "Thermal Neutron Scattering". Academic Press, London. 1965.
7. Erdelyi, A., Magnus, W., Oberhettinger, F., and Tricomi, F. G., "Tables of Integral Transforms (Bateman Manuscript Project)". McGraw-Hill, New York. 1954. (See also reference 24).
8. Jahnke, E., and Emde, F., "Tables of Functions with Formulae and Curves". Dover, New York. 1943. (See also reference 25).
9. Lanczos, C., "Discourse on Fourier Series". Oliver and Boyd, Edinburgh. 1966.
10. Landau, L. D., and Lifshitz, E. M., "Statistical Physics". Pergamon, London. 1959.
11. Lathi, B. P., "An Introduction to Random Signals and Communication Theory". Intertext, London. 1970.
12. Papoulis, A., "The Fourier Integral and its Applications". McGraw-Hill, New York. 1962.
13. Papoulis, A., "Systems and Transforms with Applications in Optics". McGraw-Hill, New York. 1968.
14. Smith, H. M., "Principles of Holography". Interscience, New York. 1969.
15. Sneddon, I. N., "Fourier Transforms". McGraw-Hill, New York. 1951.
16. Sneddon, I. N., "Fourier Series". Routledge, London. 1961.
17. Sommerfeld, A. J. W., "Optics". Academic Press, New York. 1964.

18. Stone, J. M., "Radiation and Optics". McGraw-Hill, New York. 1963.
19. Stroke, G. W., "An Introduction to Coherent Optics and Holography". Academic Press, New York. 1966.
20. Stuart, R. D., "An Introduction to Fourier Analysis". Science Paperbacks, London. 1961.
21. Titchmarsh, E. C., "Introduction to the Theory of Fourier Integrals". Clarendon Press, Oxford. 1962.
22. Watson, G. N., "A Treatise on the Theory of Bessel Functions". Cambridge University Press, Cambridge. 1962.
23. Wax, N., (editor) "Selected Papers on Noise and Stochastic Processes". Dover, New York. 1954.
24. Oberhettinger, F., "Tabellen zur Fourier Transformation". Springer-Verlag, Berlin. 1957.
25. Abramowitz, M., and Stegun, I. A., "Handbook of Mathematical Functions". Dover, New York. 1965.

Index